Is Neo-Darwinism Enough?

Is Neo-Darwinism Enough?

The Noble-Wilson Dialogue on Evolution

Denis Noble and David Sloan Wilson

Edited By James A. Barham

Library Cataloging Data

Is Neo-Darwinism Enough? by Denis Noble and David Sloan Wilson
261 pages, 6 x 9 inches
Library of Congress Control Number: 2024952931
ISBN: 979-8-9917040-2-1 (Hardcover), 979-8-89946-015-9 (Paperback),
979-8-9917040-3-8 (Kindle)
BISAC: SCI027000 SCIENCE / Life Sciences / Evolution

Book Cover Design:

Actiniae (sea anemones) from Ernst Haeckel's *Kunstformen der Natur* (Art Forms of Nature) of 1904
Image used in cover design is from Public Domain: https://commons.wikime dia.org/wiki/File:Haeckel_Actiniae.jpg

Publisher Information

Inkwell Press
2321 Sir Barton Way
Suite 140-1032
Lexington, KY 40509

Table of Contents

Editor's Preface

David Sloan Wilson and Denis Noble have long been prominent participants in the debate surrounding the interpretation of the Modern Synthesis in evolutionary biology.

More specifically, they have both frequently found themselves near or at the center of the ongoing controversy over whether the neo-Darwinian synthesis provides a reasonably complete and adequate account of biological phenomena, especially the appearance of teleology (function, purpose, goal-directedness) in virtually all biological structures and processes.

For this reason, AcademicInfluence.com invited Wilson and Noble to take part in an online Dialogue on Evolution, which we called "Is Neo-Darwinism Enough?" The point of the Dialogue was for both parties to advance the best case for their position, as well as to refute the case of their interlocutor. We are grateful that both Professor Wilson and Professor Noble accepted this invitation.

In this Dialogue, each party develops what he regards as the strongest points in favor of his own position, while also defending against what the other party alleges are its weakest points. We have suggested that each interlocutor articulate five strong points and five weak points.

In 2021, Wilson and Noble revisited their 2016 Dialogue on Evolution. At that time, each author wrote up an "Afterword," in which he discussed his reflections on the 2016 Dialogue, together with new thoughts inspired by new work in the field of evolutionary biology during the previous five years. This book has therefore been ready to go for several years and might have been published in 2022. Unfortunately, Covid intervened and with it we were also faced with finding a suitable press, thus moving the publication date to 2025. Notwithstanding, this dialogue remains of current interest, and we are happy at long last to make it available.

Thus, altogether, Wilson and Noble each contributed (1) an Interview, (2) a Major Statement, (3) a Response, (4) a Final Reply, and (5) an Afterword, in that order.

In a nutshell, in this book Wilson and Noble argue the following theses:

> **Prof. Wilson:** *The neo-Darwinian framework provides a reasonably complete and adequate explanation of biological phenomena, including the appearance of teleology in living systems. No new theoretical breakthroughs are either necessary or to be expected. Neo-Darwinism is enough!*

> **Prof. Noble:** *The neo-Darwinian framework, however necessary it may be, does not provide a complete and adequate explanation of biological phenomena, especially teleology. New theoretical breakthroughs are both necessary and to be expected. Neo-Darwinism is not enough!*

David Sloan Wilson is SUNY Distinguished Professor in the departments of Biology and Anthropology at Binghamton University. His research interests focus on multilevel selection theory, human evolution, and differentiation within species, populations, and individuals. Wilson received his PhD from Michigan State University. He is the author of many books, including *Evolution for Everyone: How Darwin's Theory Can Change the Way We Think About Our Lives* (Delta, 2007), *Does Altruism Exist?: Culture, Genes, and the*

Welfare of Others (Yale UP, 2015), and *This View of Life: Completing the Darwinian Revolution* (Pantheon, 2019).
Website: **https://davidsloanwilson.world/**.

Denis Noble is Emeritus Professor of Cardiovascular Physiology in the Department of Physiology, Anatomy, and Genetics of the Medical Sciences Division of the University of Oxford, where he held the Burdon Sanderson Chair of Cardiovascular Physiology, 1984–2004. A pioneer in computer modeling of biological organs and systems, Noble received his PhD from University College London. He is the author of several books, including *The Music of Life: Biology Beyond Genes* (Oxford UP, 2008), *Dance to the Tune of Life: Biological Relativity* (Cambridge UP, 2017), and *Understanding Living Systems*, with Raymond Noble (Cambridge UP, 2023).
Website: **https://www.dpag.ox.ac.uk/ team/denis-noble.**

James Barham
Pella, Iowa

1. David Sloan Wilson Interview

1.1. Introduction

David Sloan Wilson is SUNY Distinguished Professor in the departments of Anthropology and Biology at Binghamton University. Co-founder of the Evolution Institute, the Binghamton Neighborhood Project, and the combined Evolutionary Studies (EvoS) program at Binghamton (which has been imitated by several other universities), Professor Wilson is the author of numerous scholarly and popular books on evolutionary theory and its implications for human society. His research has been focused on the notion of group selection as a modification to mainstream neo-Darwinian evolutionary theory and on social and cultural evolution among human beings.

Among Professor Wilson's many books are *Unto Others: The Evolution and Psychology of Unselfish Behavior,* with Elliott Sober (Harvard UP, 1998), *Darwin's Cathedral: Evolution, Religion, and the Nature of Society* (University of Chicago Press, 2002), *Evolution for Everyone: How Darwin's Theory Can Change the Way We Think About Our Lives* (Delacorte Press, 2007), *The Neighborhood Project: Using Evolution to Improve My City, One Block at a Time* (Little, Brown and Co., 2011), *Does Altruism Exist?: Culture, Genes, and the Welfare of Others* (Yale UP, 2015), and, most recently, *This View of Life: Completing the Darwinian Revolution* (Pantheon, 2019).

You may learn more about Professor Wilson's many research

endeavors from his website.[1]

James Barham

Thank you very much for agreeing to participate in this interview. This is just the first phase of a Focused Civil dialogue between you and Professor Denis Noble on the question: "Is Neo-Darwinism Enough?" Accordingly, during most of this interview we will be focusing on your work in evolutionary theory. First, however, we would like you to tell our readers a little bit about yourself: when and where you were born; your family background; your education; your career—that sort of thing.

David Sloan Wilson

I was born in Norwalk, Connecticut, in 1949. My father was the novelist Sloan Wilson, who wrote two blockbusters: *The Man in the Gray Flannel Suit* (1956) and *A Summer Place* (1958). He achieved his fame in 1956, when I was seven years old. Being the son of a famous author was formative for me. I admired him tremendously, but I also worried that I couldn't measure up to him. My boyish solution was to choose interests that he could admire but couldn't do. He was a novelist, so I would become a scientist.

My family moved around a lot and I can scarcely remember my homes and hometowns. However, we spent every summer on the north end of Lake George, New York, one of the most beautiful lakes in the world. My father's family owned a hotel on the site that was torn down after World War II, leaving 600 acres of property, about a mile of lakefront, and about two dozen cottages that were maintained as a private club called the Rogers Rock Club. Roger's Rock is a cliff that rises 600 feet out of Lake George and it was on our property. Can you imagine a more glorious way to spend every summer? I spent most of my time outdoors and became a passionate fisherman, a love that has stayed with me.

---

[1] [https://davidsloanwilson.world/about-david-sloan-wilson/.]

My parents sent me to boarding school when I was 11. Their marriage was deteriorating and according to my mother, I was following my father around saying "I'm sorry! I'm sorry!" Knowing how much I loved my summers on Lake George, they sent me to a progressive boarding school called the North Country School near Lake Placid, New York, which is still going strong. It featured outdoor activities, including a working farm that the students helped to operate. I thrived there and it cemented my lifelong attachment to nature. I wear a denim barn jacket, not a gray flannel suit!

After graduating from the North Country School in the eighth grade, I tried to pick another boarding school that was just like it: the Woodstock Country School in South Woodstock, Vermont. It turned out to be a rather cruel social environment in which the students formed into cliques and the rules were too lax to keep them under control. There wasn't physical violence, but I observed a lot of psychological stress. Also, this was the 1960s and drugs were making the scene. I wasn't unduly affected, but it caused me to become somewhat of a loner, which suited my temperament anyway. I did not mourn when the school eventually went belly up.

When it came time to apply to college, I aimed high—Harvard, Yale, and Princeton—only to be rejected by all three. I ended up going to my "safety" college, the University of Rochester, which provided a fine education. I earned my PhD at Michigan State University in 1975, followed by a succession of postdoctoral and tenured faculty positions, ending up at Binghamton University[2] in 1989.

James Barham

Looking back, what would you say most influenced your decision to become an evolutionary biologist?

---

[2] Formerly, SUNY Binghamton—Ed.

David Sloan Wilson

When I decided to become a scientist at an early age, I had the white-coated lab scientist in mind. Gradually, I became aware that it was possible to become an ecologist and study organisms in their natural environment in addition to the laboratory. As you can imagine, given my love of nature, it was a very easy decision to become an ecologist. One of the great things about Rochester was the ability to work in the labs of the professors, more like a graduate student than an undergraduate. I started to function in this mode as a sophomore, thanks to an ecologist named Conrad Istock, who became my first mentor.

I was lucky to receive my higher education at a time when the biological sciences were becoming conceptually unified by evolutionary theory. I learned from Conrad that you can't be an ecologist without also being an evolutionary biologist, because all species are a product of natural selection. I was a poor student when it came to the reductionistic branches of biology, but "natural selection thinking" came very easily to me. I remember reading Ernst Mayr's 800-page *Animal Species and Evolution* (Harvard UP/Belknap Press, 1963) in a week, as if it were a novel.

Here is how I experienced conceptual unification as a student. Conrad was studying life history evolution in insects in his laboratory, but I became interested in vertical migration in zooplankton. No problem. In both cases, it was a matter of reasoning about what traits might evolve in a given species occupying a given environment. The same reasoning made it easy to switch from studying vertical migration in zooplankton to selective feeding in zooplankton—same organism, different set of traits. In graduate school I had the opportunity to study selective feeding in ant lions—different organism, same set of traits. Then I learned how to construct mathematical models of selective feeding—what would evolve in any species, given certain selective pressures. Natural selection thinking was the opposite of the old joke about experts learning more and more about less and less until they knew everything about nothing. Natural selection thinking was a passport to the study of all subjects. That's the kind of higher education in biology that I and my peers received in the 1970s: the decade that witnessed Theodosius Dobzhansky's famous dictum "Nothing in biology makes sense except in the light of evolution"[3] in 1973 and the publication of E.O. Wilson's Sociobiology (Harvard UP/Belknap Press) in 1975.

The thesis of Sociobiology was that a single theoretical framework can explain social behaviors in all species, from microbes to humans. It was celebrated as a triumph—except for the final chapter on humans, which created a storm of controversy. I was at Harvard on my first postdoctoral position in 1975, working with the ecologist Tom Schoener on the adequacy of body size as a niche difference. I read a manuscript copy of Sociobiology that Ed Wilson had placed in the library of the Museum of Comparative Zoology before it was published. I regarded the final chapter as uncontroversial and was surprised when it created such a fuss.

Studying humanity from an evolutionary perspective had a special attraction for me because it is similar to the novelistic enterprise of trying to understand the human condition. My father did this by seeing through the lens of his personal experience. I could do it by seeing through the lens of evolutionary theory. I sometimes describe myself as a novelist trapped inside the body of a scientist!

James Barham

You are best known for two things: (1) for your work on the conceptual foundations of the theory of natural selection, especially on the so-called "hierarchical model" of the selection (which we will ask you to explain shortly); and (2) for your public advocacy of evolutionary theory as a way of understanding human nature and human society, and even as a means of practical problem-solving in the social and political spheres.

---

[3] T. Dobzhansky, "Nothing in Biology Makes Sense Except in the Light of Evolution," *American Biology Teacher*, 1973, **35**(3): 125–129.

In connection with the second interest, you have been heavily engaged in institution-building. You are best-known for three initiatives, in particular: The EvoS curriculum, the Evolution Institute, and the Neighborhood Project. Could you please tell us a little bit about the nature and aims of each of these remarkable undertakings?

David Sloan Wilson
As background, it is important to establish that the study of study of evolution in relation to human affairs has a very different history than the study of evolution in the biological sciences. The latter developed more or less continuously since Darwin, while the former became politically incorrect early in the twentieth century. This is why the final chapter of Sociobiology created such a fuss. Terms such as "Evolutionary Psychology" and "Evolutionary Anthropology" weren't coined until the 1980s, signifying a renewed effort to rethink the human-related academic disciplines from an evolutionary perspective, and even these had an air of scandal about them. As a result, evolutionary training in higher education is very largely restricted to the biological sciences. If you're not a biology or a physical anthropology major, you won't learn much about evolution, and that's the way it has been since before the professors in the human-related departments were born. This problem exists around the world, unlike religious creationism, which is largely (although not entirely) confined to the United States.

As you might guess from my life story, I am among those who have been fearlessly and joyously expanding evolutionary theory beyond the biological sciences to include all things human. I regard humans as just another species that my evolutionary passport qualifies me to study. Over the years I have studied human-related topics as diverse as altruism, epistemology, personality, language, Machiavellianism, decision-making, gossip, physical attractiveness, emotion, and religion. These studies took place in parallel with my research on insects, fish, birds, parasites, and microbes.

The institution-building phase of my career began in 2003. I could see that historians would look back upon the twenty-first century as a period of synthesis of knowledge about humanity, comparable to the synthesis of biological knowledge during the Twentieth Century (which continues). Given the conservatism of academic culture, however, I could also see that without a concerted effort, decades would be required for the twenty-first century synthesis to be reflected in higher education, on my campus or any other campus. I therefore worked to create a campus-wide program for teaching evolution across the curriculum that we called EvoS (for Evolutionary Studies and pronounced as one word).

We built EvoS from elements that exist at most colleges and universities, including a scattering of faculty who are already employing an evolutionary perspective across human-related disciplines, but who are isolated within their respective departments. My book Evolution for Everyone (Delacourt, 2007) was written on the basis of EvoS and numerous academic articles describe the program in detail.[4] A sister program was started at SUNY New Paltz by the evolutionary psychologist Glenn Geher. Together we wrote and received a National Science Foundation (NSF) grant to develop our programs and create a multi-institution consortium. There is now a consortium website and online journal, about six full-fledged campus-wide programs, more nascent programs, and many more single courses that use Evolution for Everyone as a text. That's pretty good, but there is still a long way to go.

The Binghamton Neighborhood Project uses the community surrounding Binghamton University as a "field site," as an evolutionary ecologist would use the term. You can't get started studying a species from an evolutionary perspective unless you know something about

---

[4] E.g., D.S. Wilson, "Evolution for Everyone: How to Increase Acceptance of, Interest in, and Knowledge about Evolution," *Public Library of Science (PLoS) Biology*, 2005, **3**: 1001–1008; and D.S. Wilson, G. Geher, J. Waldo, and R. Chang, "The EvoS Consortium: Catalyzing Evolutionary Training in Higher Education (Introductory article to a special issue devoted to EvoS," *Evolution: Education and Outreach*, 2011, **4**(1): 8–10.

the species in relation to its environment. That's why field research is the starting point of evolutionary inquiry and all laboratory research needs to be informed by field research. Yet, most research in the human behavioral sciences is not like this and the most field-oriented disciplines, such as sociology and cultural anthropology, have also historically been the most phobic about evolution. Thus, doing what comes naturally as an evolutionary ecologist turns out to be a new model for basic and applied community-based research on humans, as I describe in my book The Neighborhood Project (Little, Brown, 2011).

A year after I started to do community-based research, I was approached by a retired political scientist and lifelong humanist named Jerry Lieberman with the idea of creating a think tank. Jerry wanted it to be science-based and was persuaded by my book Evolution for Everyone that it should be informed by evolutionary theory. I was already hooked on the idea of applying evolutionary theory to the solution of real-world problems, so I eagerly accepted Jerry's invitation and the Evolution Institute (EI) was born.

I'll have the opportunity to describe EI projects later in this interview, but here I will mention its communication outlets: This View of Life (for a general audience), the Social Evolution Forum (for a professional audience), Evonomics.com (for an economics audience), and *PROSOCIAL Magazine* (for the community of groups using PROSOCIAL, which I will describe later). Collectively, they reach hundreds of thousands of readers every month, spreading the word about the twenty-first-century synthesis in the same way as EvoS.

James Barham
Now, let us move on to your work on the foundations of evolutionary theory.

From the publication of the *Origin of Species* in 1859, to the rediscovery of Mendel's work on inheritance by Hugo de Vries, William Bateson, and others around the turn of the twentieth century, to the further elucidation of genetic principles by T.H. Morgan in his "fly room" at Columbia in the 1920s, to the "modern synthesis" of natural selection with population genetics by R.A. Fisher, Sewall Wright, J.B.S. Haldane, and others in the 1940s, the theory of evolution had always revolved primarily (with some exceptions) around individual organisms.

Darwin's key idea—supplemented and reinforced, but not essentially changed by the discoveries in genetics—was that random genetic variation at the genetic level leads to differential fitness at the phenotypic level of individual organisms, which in turn drives differential reproduction and thus change in the distribution of genes within the population to which the individuals belong.

Against this background, the idea for which you are most famous —that natural selection takes place at more than one level (is "hierarchical"), including that of the social group ("group selection") —assumes a radical aspect.

One of the main phenomena that the theory of group selection was developed to explain is what biologists call "altruism," a term of art meaning a phenotypic trait "that contributes to group advantage at the expense of disadvantage to itself."[5] Here, "advantage/disadvantage" is understood to mean Darwinian reproductive success/failure (increase/decrease in the proportional representation of an organism's genes in subsequent generations). An example would be sterile castes among social insects, such as soldier ants, worker bees, and the like.

This phenomenon—organisms which seem to act for the good of the group to which they belong, rather than their own good—seemed very difficult to account for along standard Darwinian lines which had always emphasized relative fitnesses and competition among individuals. This is the problem (and the historical context), as we understand it, of the notion of "group selection" that you have advanced in your work.

---

[5] S. Wright, "Genic and Organismic Evolution," *Evolution*, 1980, **34**: 825–843; p. 841.

Assuming little more knowledge of evolution on the part of our readers than has been sketched above, could you please explain to us in simple terms how you view the "hierarchical" nature of natural selection? That is, in a nutshell, how would you characterize your own main theoretical contribution to evolutionary theory?

David Sloan Wilson

Let me revise your historical account a little bit. According to the Christian worldview, a universe created by an all-powerful and beneficent God must be harmonious from top to bottom, from the smallest insect to the stars in heaven. The first Enlightenment thinkers, such as Isaac Newton, thought that science and reason would affirm scripture in this respect. Darwin's theory led to a very different conclusion: that the kind of functional organization that we associate with a human implement such as a watch or a single organism such as an insect, might cease to exist at larger scales, such as an animal society or a multi-species ecosystem.

Darwin was led to this conclusion by considering the evolution of traits that are "for the good of the group" and therefore morally praiseworthy in human terms. He could see that acting for the benefit of others or one's group as a whole typically placed the "altruist" at a selective disadvantage, compared to more selfish individuals within the same group. This was a problem of the first rank.

Not only did Darwin clearly see the problem, but he also saw the outline of a solution. Groups of altruists have a selective advantage over groups of selfish individuals, as surely as selfish individuals have an advantage over altruistic individuals within groups. As Edward O. Wilson and I put it in our 2007 review article titled "Rethinking the Theoretical Foundation of Sociobiology,"[6] selfishness beats altruism within groups, altruistic groups beat selfish groups, and everything else is commentary. Thus, the basic concept of multi-

---

[6] D.S. Wilson and E.O. Wilson, "Rethinking the Theoretical Foundation of Sociobiology," *Quarterly Review of Biology*, 2007, **82**: 327–348.

level selection theory was present at the start of evolutionary theory and was required to address a fundamental problem concerning levels of functional organization.

Notice that in the basic scenario imagined by Darwin, the evolving traits (altruistic and selfish behaviors) can be measured in individuals. In a genetic model, they are both coded by genes. Multilevel selection theory is concerned with *fitness differences in a nested hierarchy of scales*—favoring selfish traits (and genes) at the scale of individuals within groups and favoring altruistic traits (and genes) at the scale of groups within a multi-group population. This will become important when we consider the tortuous history of the group selection controversy.

James Barham

The idea of group selection was first brought into the mainstream of modern evolutionary discourse by V.C. Wynne-Edwards in his classic work, *Animal Dispersion in Relation to Social Behaviour* (Edinburgh: Oliver & Boyd, 1962). After much rancorous debate, Wynne-Edwards's work was considered to have been thoroughly "debunked," and was cast into the outer darkness by the neo-Darwinian mainstream. Your work has had a crucial impact on the grudging tolerance now accorded to group selection today.

Could you please tell us (1) how you became convinced that Wynne-Edwards was at least partly right; and (2) without going into too much technical detail, how your own view of group selection differs from his?

David Sloan Wilson

I wish it was otherwise, but scientists are as prone to constructing simplified patriotic histories as other people. Good scholarship is required to set the record straight. Fortunately, this period in the history of evolutionary thought is beginning to get the attention it deserves. The brief account that Elliott Sober and I gave in *Unto Others* (Harvard UP, 1998) has withstood the test of time, and has been supplemented by books such as Mark Borrello's *Evolutionary Restraints* (University of Chicago Press, 2010) and Oren Harman's *The Price of Altruism* (Norton, 2010). Here is a thumbnail history.

As important as the altruism question was, it was eclipsed by even more important questions such as the nature of heredity during the first half of the twentieth century. R.A. Fisher, J.B.S. Haldane, and Sewall Wright, the three main architects of population genetics theory, each considered the problem briefly, sketching simple mathematical models along the lines of Darwin's verbal scenario. In the meantime, many empirical biologists naively assumed that adaptations can evolve at any level of the biological hierarchy without requiring special conditions. When the need for higher-level selection was recognized, it was often assumed that higher-level selection easily trumps lower-level selection. This position became known as "naïve group selectionism" and it was a genuine problem recognized by George C. Williams in the 1950's. Williams was reacting to biologists such as W.C. Allee at the University of Chicago and he started writing his classic book, *Adaptation and Natural Selection* (Princeton UP, 1966), years before Wynne-Edwards's book was published.

Wynne-Edwards proposed that animals evolve to regulate their population size to avoid overexploiting their resources. He was aware that his hypothesis required group selection—in this sense he was not naïve—but he assumed without justification that between-group selection easily trumps within-group selection. He was not a theoretical biologist and cited Sewall Wright for support, even though Wright had written little on the subject (not to be confused with Wright's shifting balance theory, which addresses a different set of issues). Wynne-Edwards then provided an encyclopedia of examples across the animal kingdom that he interpreted in support of his hypothesis.

The reaction to Wynne-Edwards's book in 1962, along with the publication of *Adaptation and Natural Selection* in 1966, caused multilevel selection theory to occupy center stage in evolutionary biology for the first time. Note that this is about two decades after the so-called Modern Synthesis. New mathematical and computer simulation models were constructed, which suggested that the conditions for between-group selection prevailing against within-group selection were quite restrictive. It was easy to poke holes in Wynne-Edwards's empirical examples. And other ways to explain the evolution of apparently altruistic behaviors "without invoking group selection" became available, as we will shortly discuss.

Here is how I entered the field. I was offended by the idea that altruism can't evolve when I encountered it at the University of Rochester as an undergraduate student. One of Wynne-Edwards's putative examples of group selection was vertical migration in zooplankton. A common pattern is for adults to migrate during the day, while the young remain near the surface. This is probably a response to size-dependent predation, but Wynne-Edwards interpreted it as a form of mass parental care—adults migrating so they don't compete with their offspring for food. I was studying vertical migration in zooplankton and thought that Wynne-Edwards's hypothesis could not be dismissed out of hand. Far from homogenizing zooplankton populations, ocean currents and wave action concentrate them into patches. By vertically migrating, adults were very likely to horizontally separate themselves from their offspring. I included this speculation in my undergraduate thesis, but went on to other topics in graduate school.

In my final year of graduate school, a new article on vertical migration prompted me to dust off my old idea. By then I had picked up some theoretical skills and the model of within- and between-group selection that I developed went far beyond vertical migration in zooplankton. My main innovation was to define a group as the set of individuals influenced by the expression of a given trait. In fact, this is how groups are usually defined in natural language. When we say "my family," "my class," "my church," "my bowling club," we are defining groups in terms of relevant activities. In any model of social evolution, it is necessary to determine the fitness of individuals, which requires defining the set of neighbors influencing a given individual's fitness. Thus, my concept of trait-groups made explicit something that is implicitly assumed in all models of social evolution.

My model showed that while between-group selection did not invariably trump within-group selection, neither could it be categorically dismissed. The relative importance of within- and between-group selection needed to be determined on a case-by-case basis. I knew immediately that this was—or should be—a game-changer for the group selection controversy. I was so fired up that I wrote to the great E.O. Wilson requesting a meeting, in hope that he would sponsor an article for the *Proceedings of the National Academy of Sciences (PNAS)*. On my way to Harvard to meet with Wilson, I made a side trip to Stony Brook to meet with George C. Williams. My first words after striding into his office were "I'm going to convince you about group selection." His first response was to offer me a post-doc on the spot! I had other plans, but we became good friends despite our very different positions on group selection. This is how science should be.

Ed Wilson did sponsor my article for *PNAS*, which was the beginning of my relationship with him. My next problem was what to do for my PhD. My previous research topic no longer held any interest for me! My thesis adviser was a free-spirited aquatic ecologist named Don Hall. He reasoned that if my article on group selection was good enough for Ed Wilson and *PNAS*, it was good enough to be a PhD thesis. I therefore probably have the shortest PhD thesis in the history of evolutionary biology (11 pages), although I did write several other publications on other topics as a graduate student that were not included in my thesis.

James Barham

We gather that Wynne-Edwards was something of an isolated, even tragic, figure. Did you know him personally? If so, what was he like? What do you think his place in the history of science will be in the long run.

David Sloan Wilson

I never met him, but we corresponded on numerous occasions. He was a vigorous man and I don't think that he was crushed by the controversy. I regard him as less pivotal than the simplified history makes him out to be. He did not play any role in the construction and testing of theoretical models, which was the main action. His second book, published in 1986,[7] did not reflect much understanding of everything that had happened to revive multilevel selection theory, which we will shortly discuss.

If we turn our attention away from the person and focus on his hypothesis, there is now solid evidence that some consumer species do regulate their population size to avoid overexploiting their resources. This is not invariably the case, but it is sometimes the case. The best examples are parasites and diseases that reduce their virulence to avoid killing their hosts, which were showcased—and correctly interpreted—by none other than George C. Williams when he started to pioneer the subject of "Darwinian medicine." Thus, the categorical rejection of Wynne-Edwards's hypothesis does not reflect well upon his critics.

James Barham

One of the main reasons why group selection fell from favor in the '60s was the development of alternative, individual-selection-based theories to explain altruism: especially W.D. Hamilton's notion of "kin selection"[8] and Robert Trivers's concept of "reciprocal altruism" (based on game theory).[9] George C. Williams famously argued that, given the existence of these ideas, it is more parsimonious to reject group selection.

---

[7] V.C. Wynne-Edwards, *Evolution through Group Selection*. Oxford: Blackwell Scientific, 1986.

[8] R.L. Trivers, *Social Evolution*. Menlo Park, CA: Benjamin/Cummings, 1985.

[9] W.D. Hamilton, *Narrow Roads of Gene Land, Vol. 1: Evolution of Social Behaviour.* Oxford: Oxford University Press, 1996. (Contains the seminal papers on kin-selection theory from the early 1960s—Ed.)

As we understand it, your position is that kin selection and game-theoretic reciprocity are forms of group selection, so Williams was wrong. We gather that this claim remains highly controversial.

Could you briefly (if such a thing is possible!) give us your reasons for believing this? Where, exactly, did Williams go wrong?

David Sloan Wilson

George got some things very right and other things very wrong. He was right to forcefully assert the logic of multilevel selection, which is: Group-level adaptations require a process of group-level selection and should never be invoked otherwise. He was wrong to conclude on the basis of empirical evidence that between-group selection is invariably weak, compared to within-group selection. Once we abandon this sweeping empirical claim, we arrive at the reasonable conclusion that the balance between levels of selection needs to be determined on a case-by-case basis.

I know from my conversations with George that he wrote Adaptation and Natural Selection to educate a broad biological audience in basic concepts from population genetics. One of these is the concept of average effects, which is essentially the "gene's eye view" of evolution further popularized by Richard Dawkins. Consider two alleles, A and a, at a single genetic locus. They exist in three genotypic combinations, AA, Aa, and aa. They also exist in different combinations with genes at other loci and in different social groupings in a multi-group population. It is possible to average the fitness of the two alleles across all of these contexts to achieve a bottom-line estimate of which gene has the highest fitness and therefore increases in frequency in the total population. Williams called this an "accounting method" and it is indeed one of the most useful tools in the population genetics toolbox.

A problem arises, however, when average effects are taken to be an argument against group selection. When altruism evolves in a group selection model, the gene for altruism is more fit than the gene for selfishness, all things considered. It achieved its advantage on the strength of fitness differences between groups and despite opposing fitness differences within groups. The fact that it has the highest average effect, and therefore increases in frequency in the total population, is not an argument against group selection! Yet that is exactly how "the gene's eye view" of evolution became interpreted. This confusion began with Williams and was amplified by Dawkins.

James Barham

As we understand it, you believe that one of the main reasons the idea of group selection has been so maligned over the years has to do with something called the "averaging fallacy." Could you explain in simple terms what this means?

David Sloan Wilson

The concept of average effects that I just described is an example of the averaging fallacy. By itself it is a useful accounting method, but it becomes a fallacy when used as an argument against group selection. N-person game theory provides another example. Individuals employing different social strategies are assumed to interact in groups of size N. Cooperators are less fit than defectors within any given group, but groups with more cooperators are more fit than groups with fewer cooperators. The logic of multilevel selection is there for anyone to see, but it is obscured when average payoffs are calculated for each strategy, similar to the average effects of genes. Once again, there is nothing wrong with this procedure; it only becomes a fallacy when taken as an argument against between-group selection.

In this fashion, every theory of social behavior that was developed as an alternative to group selection, including kin selection theory (=inclusive fitness theory), evolutionary game theory (including reciprocal altruism), and selfish gene theory, turns out to include the logic of multilevel selection theory within its own structure. They all assume that social interactions take place within groups that are small compared to the total population. The behaviors labeled "cooperative" or "altruistic" are selectively disadvantageous within these groups and evolve only by virtue of fitness differences between groups. My concept of trait-groups helped to make this clear, along with a statistical method for partitioning selection into within- and between-group selection that was developed by George Price and convinced W.D. Hamilton that kin selection is a type of group selection. Hamilton announced his "conversion" in an article written in 1975, the same year that my *PNAS* article was published. Oren Harman's *The Price of Altruism* provides a book-length account.

So, all major theories of social evolution include the basic logic of multilevel selection and their differences are primarily a matter of perspective. This has become known as "Equivalence" and it has become accepted by most authors of peer-reviewed articles. In 1975, it was almost mandatory to say that one's ideas did not invoke group selection. Today, it is almost mandatory to say something like this:[10]

"In earlier debates, biologists tended to regard kin and multilevel selection as rival empirical hypotheses, but many contemporary biologists regard them as ultimately equivalent, on the grounds that gene frequency change can be correctly computed using either approach. Although dissenters from this Equivalence claim can be found, the majority of social evolutionists appear to endorse it."

That's quite a change! It demonstrates that the group selection controversy is essentially over. In my latest book but one, *Does Altruism Exist?* (Yale UP, 2015), I say that I offer a "post-resolution" account.

James Barham

Stephen Jay Gould famously believed that the Modern Synthesis had undergone a "hardening" after World War II, which led to a decrease in healthy diversity of viewpoints within evolutionary biology and a

---

[10] J. Birch and S. Okasha, "Kin Selection and Its Critics," *BioScience*, 2014, **65**(1): 22–32; p. 28.

growing intolerance for heterodox opinions such as yours.[11]

Thankfully, that appears to be changing now. As the late theoretical biologist Robert Rosen put it not long before his death,[12] there is

> ... a real sea change in science today; a general increase in conceptual temperature which is liquefying outmoded doctrines which have hovered around absolute zero for the past half-century or more.

Would you agree with Gould that the "hardening of the Modern Synthesis" was real? If so, what role do you believe it played in the reception of Wynne-Edwards's and your own work?

Do you agree with Rosen and us that the situation has changed drastically over the past few years, for the better?

David Sloan Wilson

I think that we need to take a nuanced view of so-called "hardening." Consider the sensory organs of any species. They have evolved to be highly selective at perceiving and processing information from the environment. In the case of our species, we can see only a narrow segment of the light spectrum and hear only a narrow segment of the sound spectrum, and we can't perceive electrical currents or magnetic forces at all. It is necessary to be selective, because attending to everything would lead to paralysis. The same is true for scientific ideas, which means that there is a positive side to "hardening." Putting some things in the foreground and others in the background can be very useful. That said, it is also necessary to periodically revisit and reconfigure such "hardenings" and there is a tendency for them to become frozen into rigid dogmas. Major "hardenings" in the history of evolutionary thought include the distinction between "Darwinian" and "Larmarckian" inheritance, the ideas associated with the Modern Synthesis in the 1940s, and the consensus against group selection in the 1960s. In each case, I agree with Gould and

---

[11] S.J. Gould, "The Hardening of the Modern Synthesis," in M. Grene, ed., *Dimensions of Darwinism: Themes and Counterthemes in Twentieth-Century Evolutionary Theory.* Cambridge: Cambridge University Press, 1983; pp. 71–93.

[12] Personal communication: Letter sent by Rosen to the Editor in 1996.

Rosen that a major loosening was in order and that decades shouldn't be required. Really, the group selection controversy should have been settled by the 1980s and the fact that it took longer does not reflect well upon the scientific process as actually practiced.

The term "Extended Evolutionary Synthesis (EES)" describes current efforts to loosen past hardenings.[13] I am an enthusiastic proponent of the EES and the This View of Life website is devoting a series of articles to it, beginning with this interview with Kevin Laland, who is heading a major grant from the John Templeton Foundation to fund research on the EES.

James Barham

As everybody knows, Richard Dawkins popularized Hamilton's idea that individual genes are the main, if not exclusive, units of selection in his book, *The Selfish Gene* (Oxford UP, 1976). Since you obviously believe Dawkins is mistaken on this point, how do you account for the fantastic popularity of his books and of the idea of the "selfish gene" (genic selection), in particular?

David Sloan Wilson

First, I would like to praise Richard Dawkins in many respects. He's a gifted writer and thinker who has turned legions of people on to evolutionary theory. He's right about lots of things, even if he's wrong about group selection. One way to rephrase your question is to ask if the "everything is caused by our selfish genes" element of Dawkins's thought is responsible for some of his popularity. The answer is almost certainly yes—in part because of its shock value (a

---

[13] This term was popularized by a meeting of 16 of the world's top theoretical biologists at the Konrad Lorenz Institute in the Vienna suburb of Altenberg in July of 2008. Many of the papers presented at this conference were later published as M. Pigliucci and G.B. Müller, eds., *Evolution: The Extended Synthesis*. Cambridge, MA: MIT Press, 2010. See, also, the popular account of the meeting in S. Mazur, *The Altenberg 16: An Exposé of the Evolution Industry.* Berkeley, CA: North Atlantic Books/Wellington, New Zealand: Scoop Media, 2010—Ed.

smart rhetorical ploy) and in part because it resonates with a broader individualistic worldview that we will be discussing later.

While on the subject of Dawkins, I am much more critical of his stance on religion, which has seriously tarnished his reputation.

James Barham

The units/levels of selection debate is a highly convoluted and controversial affair.[14] In some ways, it seems a lot closer to philosophy than to ordinary empirical science.

If your hierarchical theory is correct, then selection takes place at multiple levels. But how can we determine which units at which levels are the actual basis for selection in a given case? Is there really any real empirical substance to the units of selection debate?

David Sloan Wilson

It is sad that this question has to be asked. When we operate within the framework of MLS theory, measuring fitness differences within and between groups is ordinary empirical science. The work of Omar Eldakar (my former graduate student) on sexual conflict in water striders provides an example.[15]

Males differ greatly in their aggressiveness toward females. Omar created pools containing six males and six females, with the proportion of aggressive males varying from 100% to 0% and various mixes in between. Within every pool containing both types, the aggressive males copulated with the females more than the docile males, which was perfectly easy to quantify. However, females laid more eggs in

---

[14] See, for example, E.A. Lloyd, "Units and Levels of Selection," in D.L. Hull and M. Ruse, eds., *The Cambridge Companion to the Philosophy of Biology*. Cambridge: Cambridge University Press, 2007; pp. 44–65.

[15] O.T. Eldakar, M.J. Dlugos, G.P. Holt, D.S. Wilson, and J.W. Pepper, "Population Structure Influences Sexual Conflict in Wild Populations of Water Striders," *Science*, 2009, **326**(null): 816; and O.T. Eldakar, D.S. Wilson, M.J. Dlugos, and J.W. Pepper, "The Role of Multilevel Selection in the Evolution of Sexual Conflict in the Water Strider *Aquarius remigis*," *Evolution*, 2010, **64**(11): 3183–3189.

pools with docile males than pools with aggressive males because they weren't being harassed. This was also easy to quantify.

In a second experiment, Omar allowed free movement between the pools. Females that entered a pool with aggressive males left as soon as they could. Aggressive males were free to leave also, but the result of everyone moving was a considerable degree of clustering of females around docile males. Thus, free movement creates the variation among groups necessary for group selection to act. This was also easy to quantify.

To summarize: Selfishness beats altruism within groups (favoring aggressive males); altruistic groups beat selfish groups (favoring docile males); and in this case the balance between levels of selection maintains both types of male in the population. All of this can be quantified and many other examples can be cited.

What gives the levels of selection debate its "convoluted and controversial" aspect, as you put it, is the confusion between theories that invoke different causal processes, such that one can be right and the other wrong, and theories that describe the same causal processes from different perspectives and deserve to coexist to the extent that they provide useful insights. The concept of Equivalence does a good job of resolving this confusion. Once we become clear on Equivalence, we can do normal science within each framework and become adept at translating among frameworks.

James Barham

Thank you. That was extremely helpful.

Some have claimed that where one stands on the group selection controversy is not unrelated to where one stands politically—to put it crudely, that "neo-liberals" love old-fashioned individualistic natural selection, while communitarians and socialists prefer group selection. Do you think there is anything to this?

David Sloan Wilson

Yes, indeed. Just like other people, scientists have worldviews that influence what they find plausible and would like to be true. That's why it is important for a scientific community to include a diversity of worldviews. As long as personal biases lead to testable hypotheses, they become grist for the scientific mill.

Cultural biases are more difficult to correct when they are shared by most members of a scientific community. With the benefit of hindsight, we can see the imprint that Victorian culture had on Darwin and most of his peers on topics such as the mental inferiority of women and the superiority of European culture. No one questioned these assumptions in their everyday lives, so they had no reason to question them as scientists. I think that the swing toward individualism in evolutionary theory during the second half of the twentieth century needs to be seen against the background of a more general cultural swing that included a position known as methodological individualism in the social sciences, and of course neoclassical economic theory. When UK Prime Minister Margaret Thatcher said, "There is no such thing as society, only individuals and families," she was reflecting this trend. I hope that historians and sociologists of science pay close attention to this period.

James Barham

Let's shift gears a bit now and talk about how evolutionary theory impacts (or ought to impact) human self-understanding.

The landmark work in this area, of course, was E.O. Wilson's *Sociobiology.* In that book, if memory serves, Wilson explicitly embraced Hamilton's kin selection and abjured Wynne-Edwards's group selection. And yet very recently, he has signed on as co-author with you and others in a series of controversial papers intended to make group theory intellectually respectable once again. Has he ever told you what made him change his mind?

David Sloan Wilson

Actually, I have a different impression. Ed was as supportive of group selection as anyone could be when he wrote *Sociobiology*. Of course, he also lauded Hamilton's theory at a time when its relationship with group selection wasn't even clear to Hamilton. Ed's current critique of kin selection theory with Martin Nowak is complicated and I have faulted them, along with Richard Dawkins, for failing to absorb the message of Equivalence. However, when it comes to human self-understanding, we definitely must go beyond kin selection theory.

Human evolution is all about a shift in the balance between levels of selection. In most primate societies, members of the same group cooperate to a degree, but are also each other's main rivals. To the best of our current knowledge, our ancestors managed to suppress the potential for disruptive self-serving competition within groups, so that between-group selection became the dominant evolutionary force. Almost everything that is distinctively human, including our ability to cooperate with unrelated individuals, our capacity for symbolic thought, and our ability to transmit large amounts of learned information across generations, are forms of physical and mental teamwork that required between-group selection to evolve. Books that develop this thesis include John Maynard Smith and Eörs Szathmary's *The Major Transitions in Evolution* (Oxford UP, 1995), Terrence Deacon's *The Symbolic Species* (Norton, 1997), Christopher Boehm's *Moral Origins* (Basic Books, 2012), E.O. Wilson's *The Social Conquest of Earth* (Liveright, 2012), and Peter Turchin's *Ultrasociety* (Beresta Books, 2015).

James Barham

With respect to the evolution of human social practices and institutions, your work has ranged over many topics. But let us take as an example religion, which was the focus of your book *Darwin's Cathedral* (University of Chicago Press, 2002).

How, exactly, does a "Darwinian" explanation of religion work? How does one go about shedding light on a human social phenomenon like religion by invoking evolutionary theory, in general, and group selection, in particular?

David Sloan Wilson
Religions puzzle the secular imagination because their beliefs depart so flagrantly from factual reality and result in practices that seem so wasteful. It's easy to understand why people make blankets, but why do they burn them in sacrifice to gods for whom there is no verifiable evidence? This question has two potential answers.

First, religious beliefs and practices might be just as irrational and wasteful as they seem and persist as byproducts of psychological and social processes that are useful in non-religious contexts.

Second, despite appearances, religious beliefs and practices might have their own logic and utility after all.

Émile Durkheim was an early proponent of the latter view. He famously defined religion as:[16]

> A unified system of beliefs and practices relative to sacred things ... which unite into one single moral community called a Church, all those who adhere to them.

Durkheim also stressed the importance of symbolic thought in the organization of human societies:[17]

> In all its aspects and at every moment of history, social life is only possible thanks to a vast symbolism.

Nevertheless, over a century of scholarship in the humanities and social sciences has not led to a consensus on the "secular utility" of religion, as Durkheim put it. The tradition of functionalism that he initiated peaked in the mid-twentieth century and is currently disparaged in many quarters. When I was writing *Darwin's Cathedral* at the turn of the twenty-first century, the most authoritative theory of religion was a byproduct theory inspired by economics, which held

---

[16] É. Durkheim, *The Elementary Forms of Religious Life*. New York: Free Press, 1995; p. 44. (First published in French in 1912—Ed.)

[17] *Ibid.*; p. 229.

that gods are imaginary beings that people invent to bargain with for goods that can't be had, such as rain during a drought or everlasting life.

Evolutionary biologists are accustomed to studying whether a given trait qualifies as an adaptation vs. a byproduct, the unit of selection, and so on—all questions that can be asked for cultural evolution in addition to genetic evolution. When this theoretical toolkit started to be applied to religion, it established a consensus that did not previously exist: Appearances notwithstanding, most enduring religions have an impressive degree of secular utility at the level of the religious community, much as Durkheim posited. Religions are also replete with byproducts, just as biological adaptations are, but the view of religion writ large as a byproduct has been authoritatively rejected. This is an excellent demonstration of the "added value" that evolutionary theory brings to the human social sciences.

James Barham

We would like to return now to this question of the "softening" of the Modern Synthesis that we alluded to earlier.

It seems to many observers today that evolutionary biology is in a state of flux—even, perhaps, of long-overdue reassessment. You yourself participated in a widely advertised conference that billed itself as presenting an "extended synthesis."[18] Your interlocutor, Denis Noble, is involved with a group that calls itself the "Third Way of Evolution."[19]

For the rest of this interview, we would like to begin the process of exploring what these notions really amount to—a process which will be continued in the subsequent statements to be submitted by you and Noble.

---

[18] D.S. Wilson, "Multilevel Selection and Major Transitions," in M. Pigliucci and G.B. Müller, eds., *Evolution: The Extended Synthesis*. Cambridge, MA: MIT Press, 2010; pp. 81–93. (See, also, Note 13, above—Ed.)

[19] https://www.thethirdwayofevolution.com/.

First, we would like to ask you a fundamental philosophical question—one which most practicing scientists shy away from as obscure and remote from their day-to-day concerns—but one which we believe is not only absolutely critical to our self-understanding as animals and as rational beings, but also of potential practical importance for science itself. We are thinking of the "problem of teleology."

It has been argued by Mary Jane West-Eberhard[20] and others that biological matter is inherently adaptive and that every instance of phenotypic change (e.g., due to genetic mutation) must be understood against the backdrop of active and adaptive response and compensation to the change on the part of the organism, at least if a new viable form is to result. It this is true, then while variation may be "random" at the genetic level, it is non-random and adaptive at the phenotypic level. This means that the neo-Darwinian theory of natural selection does not dispense with teleology, but rather silently presupposes it—it is the inherent adaptive capacity of the organism that is doing all the explanatory work, not the random-variation-and-selective-retention schema.

What do you say to this? Do you think there is such a thing as internal or immanent teleology (purpose, goal-directedness), conceived of as an objective feature of "the living state of matter"?

If so, do you think the neo-Darwinian framework completely accounts for it (perhaps in terms of Ernst Mayr's notion of "teleonomy,"[21] or in some other way)?

If so, what do you say to West-Eberhard and similar critics?

David Sloan Wilson

There is a lot of ground to cover here. First, an essential book on this topic is Jablonka and Lamb's *Evolution in Four Dimensions* (MIT Press, 2006). Eva Jablonka is listed as a member on the Third Way

---

[20] M.J. West-Eberhard, *Developmental Plasticity and Evolution*. Oxford: Oxford University Press, 2003.

[21] E. Mayr, "The Multiple Meanings of Teleological," in *idem, Toward a New Philosophy of Biology*. Cambridge, MA: Harvard University Press/Belknap Press, 1988, pp. 38–66.

of Evolution website. I use it as a first text in most of my courses and agree with just about everything in it. In general, I endorse the need to study evolutionary theory and complex systems theory in conjunction with each other. After all, organisms are complex systems living in environments that are also complex systems.

However, anyone who invokes complexity must provide a way to navigate through it. Merely citing dozens of factors that interact with each other is not helpful and leads to what can be called "combinatorial paralysis." In addition, there is an ironic tendency of some thinkers to treat evolutionary theory and complex systems theory in an either/or fashion, as if to say "Darwin is dead! Long live Complexity!" Stuart Kauffman's *Origins of Order* (Oxford UP, 1993) suffered from this problem. The challenge is to study the two bodies of theory in conjunction with each other—evolution operating on complex systems.

A recent essay of mine titled "Two Meanings of Complex Adaptive Systems"[22] addresses some of the confusion associated complex systems thinking. This key term lumps two very different meanings: (1) A complex system that is adaptive as a system; and (2) A complex system composed of agents that follow adaptive strategies. A social insect colony is an example of CAS1 and an ecosystem is an example of CAS2. Many treatments of Complex Adaptive Systems fail to distinguish between these two meanings or to specify the conditions required for a system to qualify as CAS1. This amounts to a form of naïve group selectionism in authors who are otherwise very sophisticated. A complex system by itself can produce a lot of pattern, but this pattern is no more likely to be functionally adapted to a given environment than a point mutation. System-level selection is required to produce a complex system that is adaptive as a system (CAS1)!

An exception is when a Complex Adaptive System is designed by a past evolutionary process to be anticipatory or itself an evolu-

---

[22] D.S. Wilson, "Two Meanings of Complex Adaptive Systems," in D.S. Wilson and A. Kirman, eds., *Complexity and Evolution: Toward a New Synthesis for Economics.* Cambridge, MA: MIT Press, 2016; pp. 31–46.

tionary process—what William H. Calvin[23] and Henry Plotkin[24] call "Darwin machines." Epigenetic inheritance systems, forms of social learning found in many species, and forms of symbolic thought that are distinctively human (the 2nd, 3rd, and 4th dimensions of evolution discussed by Jablonka and Lamb) are all Darwin machines that evolved by genetic evolution (the 1st dimension). In addition, genetic inheritance mechanisms are a product of genetic evolution—they didn't begin as sophisticated as they are now! Thus, Wagner and Altenberg's concept of "the evolution of evolvability" has merit.[25]

All of these call for a "loosening" that is represented by the term "Extended Evolutionary Synthesis" and is being competently developed by people such as Jablonka, Lamb, and Kevin Laland.[26] But notice that the term "Extended Evolutionary Synthesis" is carefully chosen to ring in the new without announcing a break in the past. Jablonka and Lamb make the same point when they say that if Lamarckian mechanisms of inheritance do exist, the basic concept of adaptation and natural selection will remain much the same (e.g., giraffes will still evolve long necks to browse tall trees).

In another recent essay of mine titled "Intentional Cultural Change,"[27] I make the point that even if we stick to the traditional view of genetic evolution as a purposeless process ("blind variation and selective retention"), it clearly results in organisms that behave in a purposeful fashion—intelligently looking for food and mates, avoiding predators, and so on. Ernst Mayr thought it was important to call this "teleonomy" rather than "teleology," but I have always

---

[23] W.H. Calvin, "The Brain as a Darwin Machine," *Nature,* 1987, **330**: 33–34.

[24] H. Plotkin, *Darwin Machines and the Nature of Knowledge.* Cambridge, MA: Harvard University Press, 1994.

[25] G.P. Wagner and L. Altenberg, "Perspective: Complex Adaptations and the Evolution of Evolvability," *Evolution,* 1996, **50**: 967–976.

[26] K.N. Laland, T. Uller, M.W. Feldman, K. Sterelny, G.B. Müller, A. Moczek, E. Jablonka, and J. Odling-Smee, "The Extended Evolutionary Synthesis: Its Structure, Assumptions, and Predictions," *Proceedings of the Royal Society of London B,* 2015, **282**: 20151019.

[27] D.S. Wilson, "Intentional Cultural Change," *Current Opinion in Psychology,* 2016, **8**: 190–193.

thought that the similarities are more interesting to focus upon than the differences. We know since James Mark Baldwin[28] that purposeful animal behaviors can double back to influence the evolutionary process in a kind of indirect Lamarckism. Some of the recent discoveries about anticipatory mutations, epigenetic systems, and developmental systems duplicate what we have long known about learning.

Human cultural evolution indubitably has an intentional component, but it also has a very substantial blind component. Moreover, intentional planning has a way of converting to blind variation when intentions collide and result in unforeseen consequences. Cultural evolution needs to become more intentional and needs to focus directly on the solution to global problems. The intentional pursuit of lower-level interests, such as individual wealth or national self-interest, will usually be disruptive at higher scales. That's the basic message of multilevel selection theory, which is profoundly antithetical to the concept of the invisible hand in economics.[29]

James Barham

West-Eberhard is only one of many investigators who are now working under the general banner of "evolutionary developmental systems" theory (or "evo-devo"). Another similar recent development is the burgeoning field of "epigenetics"—the study of modifiable and heritable non-DNA genetic markers.

In brief, proponents of evo-devo and epigenetics are renewing a former emphasis on the whole organism in evolutionary studies, while de-emphasizing to some extent the traditional role of the neo-Darwinian model of random-variation-and-selective-retention.

What is your attitude towards these recent developments?

---

[28] J.M. Baldwin, "Development and Evolution," *Philosophical Review*, 1903, **12**(4): 442–451.

[29] D.S. Wilson and J.M. Gowdy, "Human Ultrasociality and the Invisible Hand: Foundational Developments in Evolutionary Science Alter a Foundational Concept in Economics," *Journal of Bioeconomics*, 2015, **17**: 37–52.

David Sloan Wilson

A classic article by Niko Tinbergen titled "The Methods and Aims of Ethology" is worth describing in this regard.[30] Tinbergen pioneered the study of animal behavior (ethology) and shared the Nobel Prize in Medicine with Konrad Lorenz and Karl von Frisch in 1973. Before the field of ethology was established, it wasn't obvious that a behavior such as aggression could evolve in the same way as a physical trait such as a deer's antlers. In the process of describing ethology as a branch of biology, Tinbergen wrote that four questions need to be addressed for all products of evolution:

1. What is their functional basis (if any)?
2. What is their physical mechanistic basis?
3. How do they develop during the lifetime of the organism?
4. What is their historical evolution (phylogeny)?

Ever since, "Tinbergen's four questions" have been cited as a compact description of a fully rounded evolutionary approach.

Against this background, evo-devo and epigenetics provide exciting new answers to Tinbergen's "mechanism" and "development" questions. Evo-devo is especially important because development is arguably the most neglected of Tinbergen's four questions, having been largely pushed into the background by the Modern Synthesis. However, more attention to the mechanism and development questions does not require less attention to the function question! I therefore disagree with the suggestion that good old-fashioned "natural selection thinking" needs to be de-emphasized. Reasoning on the assumption that heritable phenotypic variation exists, without needing to know about the physical or developmental basis of the variation, will always be one of the most powerful tools in the evolutionary toolkit.

---

[30] N. Tinbergen, "On Aims and Methods of Ethology," *Zeitschrift für Tierpsychologie*, 1963, **20**: 410–433.

Incidentally, I recently conducted an interview with Richard Lenski[31] that provides an overview of his research on long-term evolution in *E. coli* from the perspective of Tinbergen's four questions. I titled the interview "Evolutionary Biology's Master Craftsman" to emphasize the idea that an evolutionist is like a carpenter or a plumber, who shows up on a worksite (a particular topic of interest), sizes up the job, and pulls out the appropriate tools to get the job done. The most important tools in the evolutionary toolkit are conceptual tools, not physical tools. Tinbergen's four questions did an excellent job of describing Lenski's fully rounded evolutionary approach. Whatever we mean by the "Extended Evolutionary Synthesis" or "Third Way of Evolution," it does not require a radical departure from Tinbergen's four questions.

James Barham

Some who would agree with the recent emphasis on the whole organism at the expense of neo-Darwinism's relentless preoccupation with genes and replication would say that it still does not go far enough.

After all, it seems clear that in some sense "metabolism" (shorthand for the ensemble of functions which sustain any living being in existence) is conceptually prior to replication. There are many ways to show this, but take for instance this simple thought experiment.

Somehow or other, the first living cell came into existence. Call this Process G (for "genesis"). Now, imagine a population of cells which spontaneously undergo Process G, die off, and are replaced with another population of cells brought into existence by means of Process G. Such cells would be utterly incapable of replication, hence a fortiori incapable of evolution via natural selection. But they would clearly be alive. So, evolution is not necessary for life.

Also, consider the fact that replication is a highly regulated functional system. It is empirically well-established that replication

---

[31] The interview may be accessed here. [https://web.archive.org/web/2017070509004 8/https://evolution-institute.org/article/evolutionary-biologys-master-craftsman-an-intervie w-with-richard-lenski/?source=tvol.]

in living systems could not exist without metabolism, whereas we just showed that metabolism could perfectly well exist without replication (and indeed must have done so at the origin of life).

All of this, if correct, would mean that we ought first to endeavor to understand life itself in a far more fundamental way than we presently do, before we can hope to achieve a deep understanding of the evolutionary process. Call this idea the "New Organicism."

Would you care to comment on the New Organicism?

David Sloan Wilson

It's true that there must be something that qualifies as an organism with a metabolism before evolution has something to act upon, but evolution enters the picture so early that I don't see much point in separating them. Moreover, replication is not necessarily a big deal. It can involve the proto-organism fragmenting into parts, for example. An important point is that there can be evolution without replicators. Whole systems can replicate (the concept of a hypercycle, as I understand it) without gene-like entities within the system. Most of what you describe under the title of the "New Organicism" strikes me as similar to what I already associate with the literature on the origin of life. I'm not an expert in this area, however.

James Barham

Some authors take the New Organicism still farther than West-Eberhard and the other adepts of evo-devo and epigenetics do.

For example, James A. Shapiro[32] has recently claimed that the genome is most likely a "read-write" mechanism, and not a "read-only" mechanism as usually supposed, thus upending Francis Crick's "central dogma of biology" (the idea that the causal arrow goes

---

[32] See, e.g., J.A. Shapiro, "The Basic Concept of the Read–Write Genome: Mini-Review on Cell-Mediated DNA Modification," *BioSystems*, 2016, **140**: 35–37; *idem*, "Physiology of the Read–Write Genome," *Journal of Physiology*, 2014, **592**: 2319–2341; and *idem*, "How Life Changes Itself: The Read–Write (RW) Genome," *Physics of Life Reviews*, 2013, **10**: 287–323.

uniquely from nucleic acids to proteins, and never the reverse). The proposal, as we understand it, is that bacteria (at least) are capable of modifying their own genomes in an adaptive manner, in response to their changing physiological requirements.

If well confirmed, this phenomenon (which Shapiro also refers to as "natural genetic engineering") would not only give further support to objective teleology and the New Organicism, it would basically vindicate Lamarck over Darwin.

Are you familiar with Shapiro's work? What do you say in response to his claims.

David Sloan Wilson

I am not intimately familiar with Shapiro's work but what he describes sounds plausible to me. This would be an example of a relatively blind process of evolution resulting in a system that possesses its own capacity for adaptive change. We already have a good intuition about this at the behavioral level with myriad forms of phenotypic plasticity. In theory, there is no reason why comparable mechanisms can't exist for developmental programs, replication machinery, anticipatory mutations, and so on.

On Lamarck vs. Darwin, what most people think they know is based on the same kind of simplified and patriotic history that I described for the group selection controversy. The real history is admirably summarized by Jablonka and Lamb in Evolution in Four Dimensions. To begin, Darwin was Lamarckian! More important, if anything requires loosening, it is the so-called central dogma of biology and the prohibition against thinking about Lamarckian forms of inheritance.

James Barham

In closing, we would like you to tell us—in bulleted list format, if you like—what you consider to be the five strongest arguments in support of the consensus view within evolutionary biology that "neo-

Darwinian is enough," as well as the five weakest arguments that critics of neo-Darwinism commonly advance.

David Sloan Wilson

In a less oppositional spirit, I will list 10 points that stake out the productive middle ground.

1. We should never forget the power of evolutionary theory as originally formulated by Darwin and Wallace. Organisms vary. Their differences often make a difference in terms of survival and reproduction. Offspring tend to resemble their parents. Given these three conditions, the properties of organisms change over time. They become well adapted to their environments. This will always be the centerpiece of evolutionary theory, no matter how many syntheses come and go.

2. Tinbergen's four questions concerning function, mechanism, development, and phylogeny are an admirable description of a fully rounded evolutionary perspective. The most productive research programs will address all four questions in conjunction with each other.

3. Darwin based his theory on the concept of heredity and knew nothing about genes. Yet, evolutionary science became highly gene-centric during the twentieth century, as if the only way for offspring to resemble their parents is by sharing genes. Evolutionary science needs to return to its roots by including all mechanisms of inheritance.

4. The study of evolution in relation to human affairs lags behind the study of evolution in the biological sciences by almost a century. Basic literacy in evolutionary theory needs to be taught in the human-related disciplines, in addition to developments that are new for biologists. Also, biologists must become literate about the human-related disciplines to begin studying human cultural evolution.

5. George C. Williams was right to insist in his 1966 classic, *Adaptation and Natural Selection*, that adaptation at any level of a multi-tier hierarchy requires a process of selection at that level and tends to be undermined by selection at lower levels. Williams and others were wrong to claim that higher-level selection is invariably weaker than lower-level selection. Instead, the balance between levels of selection must be evaluated on a case-by-case basis. This should be part of basic training in evolutionary science.

6. Multilevel selection is especially important for the study of evolution in relation to human affairs because we are such a group-selected species. Moreover, the solution to nearly every important problem in modern life requires expanding the scale of functional organization by managing the process of cultural multilevel selection.

7. Behaviors that qualify as intentional can evolve by a process of blind variation and selective retention that does not qualify as intentional. The intentional behaviors can then double back to influence evolutionary processes. This basic insight has many applications, including human cultural evolution and phenotypic plasticity in all organisms. Developmental programs, mechanisms of genetic replication, and mechanisms of genetic change can be "directed" in the same way as more familiar forms of phenotypic plasticity.

8. Organisms are complex systems inhabiting environments that are also complex systems. Hence, evolutionary theory needs to be studied in conjunction with complex systems theory. This is challenging, however, because the study of complexity is inherently complex. Merely invoking complexity is not helpful because it leads to combinatorial paralysis. Complex systems by themselves do not result in functional design. A process of selection is required, no less than for simple systems.

9. Scientific inquiry is itself a highly managed process of cultural evolution that works well under some circumstances and breaks down under other circumstances. The more we explicitly regard science as a cultural adaptation, the better we can make it work.

10. The concept of Equivalence, which emerged largely from the group selection controversy, is relevant to other controversies, as well. Sometimes theories are different because they invoke different causal processes, such that some can be right and others wrong. Alternatively, theories can be different by viewing the same causal processes from different perspectives, deserving to coexist to the extent that they provide useful insights. Arguing equivalent theories against each other is a waste of time. Instead, it is necessary to become literate about each theory and adept at translating from one to the other. Decades of futile debate can be saved by becoming mindful about Equivalence.

James Barham

Finally, could you tell us about your plans for the immediate future? Do you have a new book in the works? Any new projects to advance the cause of evolution?

David Sloan Wilson

Very few people realize how much evolutionary science is needed to solve the problems of modern life. That is my current passion, which I pursue largely through the auspices of the Evolution Institute. It is also the subject of my next book, titled *This View of Life: Completing the Darwinian Revolution.* It will be completed in a couple of months and published by Pantheon books in 2019.

James Barham

Thank you very much for taking the time to participate in this Interview, and in the upcoming Dialogue on Evolution with Professor Noble. We appreciate it very much indeed, and we are sure that our readers will profit greatly from your insights.

David Sloan Wilson

Thank you! Your questions have been better-informed than almost any other interview I have taken part in. Congratulations for your own literacy and for increasing the literacy of others.

2. Denis Noble Interview

2.1. Introduction

Denis Noble is Emeritus Professor of Cardiovascular Physiology in the Department of Physiology, Anatomy, and Genetics of the Medical Sciences Division of the University of Oxford. A Fellow of the Royal Society and a pioneer of systems biology, Professor Noble held the Burdon Sanderson Chair of Cardiovascular Physiology in that department from 1984 until his retirement in 2004.

Professor Noble, who in 1960 developed the first successful mathematical model simulating the activity of the living heart, is the author of over 450 scientific papers, some of the most important of which are collected in *The Selected Papers of Denis Noble CBE FRS: A Journey in Physiology Towards Enlightenment,* edited by Denis Noble, et al. (Imperial College Press, 2012). He has also co-edited a volume of essays, *The Logic of Life: The Challenge of Integrative Physiology,* with C.A.R. Boyd (Oxford UP, 1993), and has authored two well-received books aimed at a general audience, *The Music of Life: Biology Beyond the Genome* (Oxford UP, 2006) and *Dance to the Tune of Life: Biological Relativity* (Cambridge UP, 2017). Most recently, he has co-authored a volume in the distinguished Cambridge "Understanding Life" series: *Understanding Living Systems,* with Raymond Noble (Cambridge UP, 2023).

More information about Professor Noble and his work may be obtained from his website.[1]

---

[1] [https://www.denisnoble.com/.]

James Barham

Thank you very much for agreeing to participate in this Interview, as well as in the forthcoming Dialogue on the conceptual foundations of evolutionary theory with David Sloan Wilson.

Before we get to the substance of the Interview—the way in which your groundbreaking work in the physiology of the heart led you to question the reductionist consensus in biology—we would like to ask you to share with us some of your personal story.

When and where were you born? What did your parents do? What is your religious background, if any? Where did you attend school and university? How did you get interested in science? Anything you would like to share with us about your personal journey.

Denis Noble

I am a child of the 1930s, one of the darkest decades of the twentieth century. In 1936 Hitler was already well-ensconced in Nazi Germany and the sinister events that led to the Second World War were rapidly unfolding. These included the Anti-Comintern Pact between Germany and Japan, signed on 25 November just days after I was born. The Spanish Civil War had begun. Mussolini and other fascist dictators all over Europe were queuing up to join with the Nazis. Anti-Jewish propaganda was alarmingly threatening, leading to *Kristallnacht* in 1938, while one of the last *Kindertransporten* from Vienna was to bring my future PhD supervisor to Britain.

My father, George, had fought in the First World War, was wounded in the Battle of the Somme, and fought with the fledgling Royal Air Force (RAF). He knew at first-hand the terrible consequences of all-out war, most particularly for working-class people like him. He knew what it was like to be a soldier when 90 percent of your colleagues were defenseless: to be cruelly mown down by machine guns. He was lucky to be only injured. There was a narrow escape during the Second World War, too, but that part of the story comes later.

He and my mother, Ethel, worked as jobbing tailors, making suits to order for rich clients of Saville Row tailoring firms. Some of my earliest recollections are of them listening to the BBC news on the radio during the war. My father would be sewing suits while sitting cross-legged on the huge iron shelter that filled the main room, while my younger brother and I hid underneath in case the bombs fell. We also had a concrete shelter in the small garden at the back. But, fortunately, he never used that; he couldn't afford to lose time from his work. That led to the second lucky escape. Hitler's bomb made a direct hit on the garden shelter. The house was a ruin but we survived. The first photograph of me, and a younger brother, taken shortly afterwards shows two emaciated children with their ribs clearly visible—we resembled the underfed children of war-torn poverty that, sadly, the world still sees today, except for the obvious fact that we look happy. We owe the smiling faces to our parents.

Shortly after the war finished, I was sent to Emanuel School in London. In those days, poor children were given state-funded places. I was there from 1947 to 1955. I was hopeless at sports—imagine those wasted arms and legs in the middle of a rugby scrum! But an absolutely dedicated set of master's in chemistry, physics, maths, and biology fired my excitement with science. The outcome was that I went on to study medicine at University College London (UCL).

I was the first member of my family ever to go to university. In those days only five percent of the population benefited from such privilege. It felt like a privilege, too. UCL was packed with household names in the various sciences, philosophy, mathematics, and many other disciplines. The successes of the reductionist approach to biology were also very apparent. I remember seeing the first images using electron microscopy from Hugh Huxley's work showing the individual protein molecular filaments responsible for muscle contraction. And I could witness the excitement of Bernard Katz's work showing the quantal nature of neuromuscular transmission.

Sadly, just as I was beginning to experience these heights of intellectual endeavor, we suddenly became much poorer through the relatively early death of my father. We became a single-parent family, and there were three younger brothers to bring up. Paying the family's electricity bills from what remained of a student grant was not easy. I shared with our mother the weekly anxieties of managing on very tight budgets.

It was impossible after that tragic event to imagine how my own career would eventually develop, and even whether I could continue as a student. We couldn't see much beyond the end of each week. My life had to focus entirely around the family, the academic work, and the long cycling journeys across London between the two. With no money left over, student social life became an unnecessary luxury, but therefore no longer a distraction.

It was therefore very fortunate that I won a prestigious Bayliss-Starling scholarship providing the full funding for graduate work towards a doctorate. It became possible for me to branch out during my graduate research years. I explored what UCL could offer in physics, in mathematics, and in philosophy. Often enough I was the only biologist in these classes. There was no specifically designed course for such an omnivorously hungry student. I simply made it up as I went along, and somehow UCL gave me the freedom to do that.

This led to my being the only biologist in the whole university to beg for time on a precious early valve computer: one of the first machines used for seriously challenging mathematics, the Ferranti Mercury. They were so expensive that I doubt whether Ferranti ever made more than a dozen or so of those worldwide. I was given the worst time slot each day: 2–4 am! Somehow, in between doing experiments on the heart during the day to obtain the data, I used the evenings and early mornings to develop the mathematical skills to do computer modelling in biology. This experience played havoc with my circadian rhythms and looking back on it I don't know how it was possible to do that for days on end before crashing out for a long weekend sleep.

James Barham

An amazing story! Thank you very much for sharing it with us.

Next, we would like to hear about your early research on the heart, leading as we understand it to two ground-breaking papers published in *Nature* in 1960. Could you please describe the nature of this research and the resulting model for us, in terms suitable for a lay audience?

Denis Noble

So, why was I the only biologist working on an early electronic computer? With my supervisor, Otto Hutter (the *Kindertransporten* boy from Vienna, but now established on the faculty of UCL), I had made an important discovery by measuring electric current flow in heart cells. By manipulating the ions present in the bathing solution, we were able to distinguish two kinds of potassium channels. The first, which we naturally called iK1, resembled the phenomenon of anomalous rectification observed by Bernard Katz in skeletal muscle.[2] Instead of being activated during the action potential, the channel is rapidly inactivated. This produces electrical behavior that is best described as a rectifier. Nowadays, we call it inward rectification since the channel can pass potassium ions into the cell much more easily than out of the cell. The other type of channel we called iK2. This type resembles that discovered in nerve axons by Hodgkin and Huxley in 1952[3] in passing outward current more easily than inward current. But the cardiac channel activates very slowly indeed, around 100 times slower than in nerve.

With these experimental results I could put 2 and 2 together and hope to make 4. I could see that the slow activation would be very important in allowing the action potential in heart to be very long

---

[2] B. Katz, "Les constantes électriques de la membrane du muscle," *Archives des sciences physiologiques*, 1949, **2**: 285–299.

[3] A.L. Hodgkin and A.F. Huxley, "A Quantitative Description of Membrane Current and its Application to Conduction and Excitation in Nerve," *Journal of Physiology*, 1952, **117**: 500–544.

compared to nerve, and the inward rectifier channel would help by reducing the amount of sodium or other ions that would need to enter to maintain the action potential. The question was whether this would work quantitatively. That could be answered only by constructing a mathematical model, just as Hodgkin and Huxley did for nerve. That is what led me to beg the guardians of the Mercury computer to give me time. Initially, they didn't agree: "You don't know enough mathematics (totally correct) and you don't know how to write computer programs." Wow! I hadn't realized that would be such a problem; I somehow thought that I would just give them my equations and the computer would do the rest.

The only solution was to ask a maths lecturer to allow me to attend his course and have my assignments marked, and to buy the programming manual to learn how to program. Most of that was in machine code. Those were the days before structured computer languages like Fortran. The first maths lecture was totally incomprehensible and the programming manual was not much better. But I knew I had a great project, so I stuck it out. After a few of the lectures I started getting full marks for the assignments and within a month or two I could see how to write the programs. I went back to the guardians of Mercury. They still shook their heads in disbelief. But they relented and gave me the two hours each night on the machine.

The results exceeded my own expectations, let alone those of my supervisor. Not only was it possible to reproduce the detailed shape of the action potentials in the region of the heart that I was studying. As a bonus, a form of pacemaker rhythm was also reproduced. Some very counterintuitive resistance changes were also explained. The result was the two publications in *Nature* in 1960, even before my doctoral thesis was submitted.[4]

I have to admit, though, that while the successes of that work were impressive, there were cracks in the explanations. Many years later

---

[4] O.F. Hutter and D. Noble, "Rectifying Properties of Heart Muscle," *Nature*, 1960, **188**: 495; and D. Noble, "Cardiac Action and Pacemaker Potentials Based on the Hodgkin-Huxley Equations," *Nature*, 1960, **188**: 495–497.

cardiac electrophysiology became much more complex. The ways in which that happened are very relevant to how I became involved in evolutionary biology. So, I will leave that matter to a later question.

James Barham

You first came to our attention many years ago as the co-editor of *The Logic of Life* (Oxford UP, 1993), which contained among many other wonderful papers an important essay entitled "Self-Organizing Systems" by F. Eugene Yates (1927–2015), the longtime President of the American Physiological Society and another pioneer of systems biology.[5]

One of the reasons why we found Yates's work so fascinating was the latter's use of the mathematical formalism known as "nonlinear dynamics" (which derives from work by Henri Poincaré at the turn of the twentieth century, but only began to find application within biology in the 1960s). More specifically, the mathematical object called a "nonlinear attractor" seemed like a promising way to model the global teleological (purposive, goal-directed—see below) aspect of biological functions, while fully acknowledging their objective reality (that is, without attempting to "reduce" them to local mechanistic interactions only).

We were very sorry to hear that Yates passed away recently. Did you know him well? What was he like?

How would you describe the relationship between Yates's work on "homeodynamics" (the modeling of generalized biological functions via nonlinear dynamics) and your own work on the heart? Do you agree that the concept of a nonlinear attractor is a useful tool for helping us to think about the teleological character of living systems?

---

[5] F.E. Yates, "Self-Organizing Systems," in C.A.R. Boyd and D. Noble, eds., *The Logic of Life: The Challenge of Integrative Physiology*. Oxford: Oxford University Press, 1993; pp. 189–218. See, also, the seminal volume that Yates himself edited, *Self-Organizing Systems: The Emergence of Order*. New York: Plenum Press, 1987; as well as the principal statement of his "homeodynamics" theoretical framework, "Order and Complexity in Dynamical Systems: Homeodynamics as a Generalized Mechanics for Biology," *Mathematical and Computer Modelling*, 1994, **19**: 49–74.

Denis Noble

Sadly, I did not know Eugene Yates personally. Someone told me that we should include him as an invited author in *The Logic of Life* (OUP, 1993), and I was very pleased we did. I was also pleased with the reception of *The Logic of Life* at the 1993 International Union of Physiological Sciences (IUPS) Congress, for which it was produced. It resulted in full-page articles in *The Guardian*[6] and *The Observer*.[7]

I believe its effect was largely to reinforce the convictions of the converted. I doubt whether it was widely read outside the domain of the physiological sciences. But it is an important publication for another reason also. My own chapter in the book, written with Richard Boyd, "The Challenge of Integrative Physiology,"[8] shows that even two decades ago I was seriously doubtful about purely gene-centric explanations of biology. The more dogmatic neo-Darwinists sometimes pretend that I have no track record in the field.[9] Actually, I organized the first debate on Dawkins's *The Selfish Gene* way back in 1976. I have at least 50 publications in peer-reviewed journals that relate to genome-phenotype relations and to evolutionary biology. The more recent ones (last 10 years) are available as The Music of Life Sourcebook.[10]

To return to Eugene Yates, the concept of homeodynamics certainly fits my work. The process of heart rhythm is a non-linear attractor. It is an attractor because, within physiological ranges of initial parameters, the system of equations naturally moves towards the limit cycle that describes the rhythm. It is non-linear because

---

[6] [http://musicoflife.website/pdfs/Guardian.pdf.]

[7] [http://musicoflife.website/pdfs/Observer.pdf.]

[8] D. Noble and C.A.R. Boyd, "The Challenge of Integrative Physiology," in C.A.R. Boyd and D. Noble, eds., *The Logic of Life: The Challenge of Integrative Physiology.* Oxford: Oxford University Press, 1993; pp. 1–13.

[9] [http://musicoflife.website/Answers-jonny-cum-lately.html.]

[10] [http://www.musicoflife.website/pdfs/The%20Music%20of%20Life-sourcebook. pdf.]

perturbations in both directions show threshold behavior in which beyond a certain size they lead to run-away behavior.

I also agree that the concept of a nonlinear attractor is a useful tool for thinking about the teleological character of living systems. What do we mean by "teleology" if not the tendency of a system to move towards the function that serves its interests in the organism as a whole, i.e., to have a goal? As I will argue later in this Dialogue, that does not require us to believe that there was a creator that designed the cardiac pacemaker. The term "final cause" has unfortunately created the impression that there is some ultimate goal in the universe from which all other forms of teleology derive. By contrast it is sufficient in my view to see teleological behavior as emergent during evolution.

James Barham

We rehearsed some of the basics of the standard neo-Darwinian theory of natural selection (i.e., random variation of genes and selective retention of corresponding phenotypes) in our interview with your interlocutor, David Sloan Wilson, so there is no need to repeat that here.

Clearly, most evolutionary biologists feel that natural selection represents the only genuine theory in biology comparable to theories in physics and chemistry, while the sort of work in physiology you have done is mere filling in of details. In short, we suspect they would not see the relevance of your work to theirs.

So, why should a physiologist be concerned about the proper interpretation of the theory of evolution? What possible role could such a scientist play in such discussions?

Denis Noble

This is a good question because the neo-Darwinist Modern Synthesis is almost designed to exclude physiology from any role other than to preserve the "vehicle" that carries the genome from one generation to another. If all inherited change is random with respect to function, and if the genome really is isolated from changes in the phenotype and its environment, then physiology has no role in the evolutionary process other than to express what the genome dictates and to allow natural selection to work on the outcome.

I developed serious doubts about this way of viewing biology as a consequence of my work on cardiac rhythm. The first model I created in 1960 was very simple indeed. Just four kinds of ion channels and no representation of any intracellular controlling processes such as the way in which calcium triggers contraction. I also found that such a model represents the process of rhythm as very fragile. A knockout of any of the genes for those protein channels would cause the rhythm to stop completely.

During the next two decades many more ion channels were discovered experimentally; Dick Tsien and I greatly extended the range of potassium channels;[11] and Harald Reuter discovered the calcium channel,[12] as well as an exchange mechanism allowing sodium and calcium ions to exchange for each other.[13] During the 1970s and 1980s, therefore, we needed to incorporate all of these additional mechanisms. Dick Tsien, Eric McAllister, and I first did that for the discoveries up to 1975,[14] while Dario DiFrancesco and I did so for the discoveries up to 1985.[15]

Some physiologists even wondered why all of this complexity was necessary when nature could clearly build rhythm mechanisms

---

[11] D. Noble and R.W. Tsien, "The Kinetics and Rectifier Properties of the Slow Potassium Current in Cardiac Purkinje Fibres," *Journal of Physiology*, 1968, **195**: 185–214; D. Noble and R.W. Tsien, "Outward Membrane Currents Activated in the Plateau Range of Potentials in Cardiac Purkinje Fibres," *Journal of Physiology*, 1969, **200**: 205–231; and D. Noble and R.W. Tsien, "Reconstruction of the Repolarization Process in Cardiac Purkinje Fibres Based on Voltage Clamp Measurements of the Membrane Current," *Journal of Physiology*, 1969, **200**: 233–254.

[12] H. Reuter, "The Dependence of Slow Inward Current in Purkinje Fibres on the Extracellular Calcium Concentration," *Journal of Physiology*, 1967, **192**: 479–492.

[13] H. Reuter and N. Seitz, "The Dependence of Calcium Efflux from Cardiac Muscle on Temperature and External Ion Concentration," *Journal of Physiology*, 1968, **195**: 451–470.

[14] R.E. McAllister, D. Noble, and R.W. Tsien, "Reconstruction of the Electrical Activity of Cardiac Purkinje Fibres," *Journal of Physiology*, 1975, **251**: 1–59.

[15] D. DiFrancesco and D. Noble, "A Model of Cardiac Electrical Activity Incorporating Ionic Pumps and Concentration Changes," *Philosophical Transactions of the Royal Society B*, 1985, **307**: 353–398.

with much less, as I had shown in 1960. Critics of the methods we were using to record and distinguish between the different channels even proposed that most of them were artefacts of the recording methods. The answer to that problem is far more interesting and significant from a functional point of view. The complexity builds in robustness. Robustness is important in enabling organisms to resist unfavorable changes in their environments or genomes.

In the case of heart rhythm, we demonstrated that robustness in 1992 by showing that a protein mechanism that can contribute up to 80 percent of the ion current flow responsible for generating rhythm in the natural pacemaker of the heart, the sinus node, can be blocked completely with only a 10–15 percent change in frequency.[16] The rhythm mechanism is therefore highly robust and relatively insensitive to what is happening at the molecular level.

Block of a protein mechanism is equivalent to a gene knockout experiment. What this shows, therefore, is that knockout experiments cannot be relied on to reveal quantitative relationships between genotypes and phenotypes. The difference between 80 percent function and 10 percent function is far too great an error to be ignored. The only way in which accurate quantitative relations can be obtained is by reverse-engineering from physiological models of the interactions between all the proteins (and therefore genes) involved. But that means that we must first understand the physiology in order to build the models with which to reverse-engineer.

That brings physiology back into relevance at least in establishing genome-phenotype relations.

This problem is even more extensive than I demonstrated. In a landmark study in 2008[17] all 6000 genes in yeast were studied using

---

[16] D. Noble, J.C. Denyer, H.F. Brown, and D. DiFrancesco, "Reciprocal Role of the Inward Currents Ib,Na and If in Controlling and Stabilizing Pacemaker Frequency of Rabbit Sino-Atrial Node Cells," *Proceedings of the Royal Society B*, 1992, **250**: 199–207; and D. Noble, "Differential and Integral Views of Genetics in Computational Systems Biology," *Journal of the Royal Society Interface Focus*, 2011, **1**: 7–15.

[17] M.E. Hillenmeyer, et al., "The Chemical Genomic Portrait of Yeast: Uncovering a Phenotype for all Genes," *Science*, 2008, **320**: 362–365.

individual knockouts. Eighty percent of the knockouts were silent in the sense that no change in metabolic or reproductive activity was observed. That does not mean that those genes have no function. It means simply that the organisms are extremely well buffered against changes in their own genomes. Once the organisms were stressed by depriving them of various nutrients, it was possible to reveal that most of the 80 percent have functional roles.

At the least, such experiments and calculations force one to think deeply about the basis of gene-centric views of biology. But they do not, in themselves, challenge the neo-Darwinist view of evolution. For me, they were more a way back into thinking about evolution again, as I have done, on and off, for about 50 years.

What followed was like a falling domino cascade. Once one central issue in gene-centrism comes into question, others inevitably also come under suspicion. I will now list some of the dominoes that fell in the wake of questioning the gene-centric view of genome-phenotype relationships.

1. The Central Dogma of Molecular Biology. The idea of a one-way determinate read-out of genome sequences (if that is taken as the meaning of the Central Dogma[18]) doesn't make much sense to a physiologist. The 200 or so cell types in a vertebrate organism all have the same genome. Each cell clearly controls its genome to produce a pattern of gene expression that is unique to that type. Moreover, the environment of each cell type, formed by the tissues

---

[18] Francis Crick's statement of the Central Dogma in 1970 was:

"The central dogma of molecular biology deals with the detailed residue-by-residue transfer of sequential information. It states that such information cannot be transferred back from protein to either protein or nucleic acid." (Francis Crick, "The Central Dogma of Molecular Biology," *Nature*, 1970, **227**: 561–563; p. 561.)

I have italicized "such information" and "from protein" since it is evident that the statement does not say that no information can pass from the organism to the genome. Otherwise, it would be impossible for different cell types to construct and reproduce themselves from the same genome. The answer is that cells control the pattern of gene expression. Many different patterns can be generated from the same genome, just as many different patterns of music can be generated using the same organ pipes.

and organs the cells find themselves in, also contributes to control of the genome. Within the body, therefore, a form of Lamarckian inheritance of acquired characteristics is rampant.

2. The Weismann Barrier. If that can happen within an organism, why can't it also happen across generations? In 1998 I interacted with the renowned evolutionary biologist John Maynard Smith (1920–2004) during a Novartis Foundation Symposium on the limits of reductionism.[19] Though working entirely within the Modern Synthesis framework, he did acknowledge some of its weak points. Two quotes from his book, *Evolutionary Genetics* (Oxford UP, 1998), suffice to illustrate this.

On Lamarckism, Maynard Smith writes:[20]

[It] is not so obviously false as is sometimes made out.

On Weismann, he writes:[21]

[I]t is not clear why he thought it [the germ line is independent of changes in the soma] is true.

I entirely agree with both remarks—but they don't go nearly far enough.

Weismann's tail-cutting experiments, in which he showed that no tailless mice were ever born to mice whose tails had been amputated, only tests whether a surgical mutilation can be inherited. Half a century later, Conrad H. Waddington (1905–1975) performed the correct experiment, which was to change the environment. He treated fruit fly embryos with gentle heat or with ether and produced

---

[19] Gregory R. Bock and Jamie A. Goode, eds., *The Limits of Reductionism in Biology* (Novartis Foundation Symposium 213). Chichester, UK: John Wiley & Sons, 1998.

[20] John Maynard Smith, *Evolutionary Genetics*, 2nd ed. Oxford: Oxford University Press, 1998 p. 8. (The great French biologist and evolutionist, Jean-Baptiste Lamarck (1744–1829), is today best [if not entirely fairly] remembered for the idea of the "inheritance of acquired characteristics."—Ed.)

[21] *Ibid.* (The German biologist, August Weismann [1834–1914], gave his name to the idea that changes at the level of "soma" [all cells other than the gametes] cannot causally influence the "germ plasm" [i.e., the genes].—Ed.)

variants from which he bred. After only a few generations it was no longer necessary to give the environmental stimulus. The change had become assimilated into the genome so that the changed phenotype bred true even without the environmental stimulus.[22] Waddington coined the term "epigenetics."

3. Transgenerational Epigenetics. Modern biology has greatly extended Waddington's epigenetic idea to include many processes that were completely unknown to him. In addition to control of gene expression by transcription factors, we now include various chemical processes by which the genome sequences can be marked in a way that modifies gene expression. Both DNA itself and the histone proteins around which it is wrapped can be marked.

Initially, it was thought that these marks are always removed between generations. But we now know that is not always true.[23] This has led to the creation of the field of transgenerational epigenetics.[24] Defenders of the Modern Synthesis usually dismiss this evidence by saying that such transmission is rare and that it always dies out after a few generations. I have three answers to that.

First, there are now well-established examples where it does not die out even over many generations.[25]

Second, even if the process is rare, so is speciation. As a mechanism contributing to the rare process of speciation, rare trans-

---

[22] C.H. Waddington, "The Genetic Assimilation of the Bithorax Phenotype," *Evolution*, 1956, **10**: 1–13; D. Noble, "Conrad Waddington and the Origins of Epigenetics," *Journal of Experimental Biology*, 2015, **218**: 816–818; and C.H. Waddington, *The Strategy of the Genes*. London: Allen and Unwin, 1957 (reprinted 2014).

[23] D. Noble, "How Widespread is Trans-generational Inheritance of Acquired Characteristics?," Music of Life website. [http://musicoflife.website/Answers-transgenerational%20inheritance.html.]

[24] T. Tollefsbol, ed., *Transgenerational Epigenetics: Evidence and Debate*. Waltham, MA: Academic Press, 2014.

[25] O. Rechavi, G. Minevish, and O. Hobert, "Transgenerational Inheritance of an Acquired Small RNA-based Antiviral Response in *C. elegans*," *Cell*, 2011, **147**: 1248–1256; and V.R. Nelson, J.D. Heaney, P.J. Tesar, N.O. Davidson, and J.H. Nadeau, "Transgenerational Epigenetic Effects of Apobec1 Deficiency on Testicular Germ Cell Tumor Susceptibility and Embryonic Viability," *Proceedings of the National Academy of Sciences, USA*, 2012, **109**: E2766–E2773.

generational inheritance of epigenetic changes could clearly occur. In a recent study of Darwin's finches, this is what seems to have happened.[26] Both epigenetic (meaning DNA marking) and genetic (meaning DNA sequence) changes are correlated with the evolutionary distance between the different species. The authors of the study conclude that both interacted in the process of speciation.

Third, it is important to note that testing whether epigenetic marking dies out after a single generation exposure to the environmental change does not test for what may happen during evolutionary change since, in any realistic case, the environmental stimulus would continue to act over many generations. Waddington's idea of genetic assimilation can then occur to ensure that the change becomes more permanent.

4. Lateral transfer of DNA. The Modern Synthesis was based on Darwin's idea of the tree of life, radiating from a common ancestor. We now know that the tree is more like a network, particularly in the early branches. DNA is not just transferred vertically from generation to generation; it can also be transferred laterally between organisms, even between different species.

5. Symbiogenesis. I was given the huge privilege of interacting with Lynn Margulis (1938–2011) on an almost daily basis during the academic year 2008–9, when she was the Eastman Visiting Professor at Oxford University. It so happens that this chair is attached to my own college, Balliol, so we often met at lunch. That led to me chairing a landmark debate between her and Richard Dawkins, together with Martin Brasier (symbiosis in corals) and Stephen Bell (prokaryotes). It lasted four hours and is fully recorded online.[27]

---

[26] M.K. Skinner, C. Guerrero-Bosagna, M.M. Haque, E.E. Nilsson, J.A.H. Koops, S.A. Knutie, and D.H. Clayton, "Epigenetics and the Evolution of Darwin's Finches," Genome Biology and Evolution, 2014, 6: 1972–1989.

[27] [https://vimeo.com/51768266.]

Richard Dawkins was the better debater, judged by some audience reactions. But Lynn had killer lines. I shall never forget this interaction:

> **Dawkins**: It [neo-Darwinism] is highly plausible, it's economical, it's parsimonious, why on earth would you want to drag in symbiogenesis when it's such an unparsimonious, uneconomical [theory]?
>
> **Margulis**: Because it's there.

That's it in a nutshell. What is there, what exists, is the starting point of all science.

6. Genome reorganization (natural genetic engineering). I have also had the great privilege of interacting over several years with James Shapiro at the University of Chicago. James taught me to understand the significance of the work of Barbara McClintock (1902–1992), the discoverer of mobile genetic elements, for which she received the Nobel Prize in 1983. Neo-Darwinians argue that this is just an example of a large mutation. I argue that if large, *already-functional* sequences are moved around the genome, then potentially existing or new functions travel with the sequences. The Modern Synthesis was built on the idea of the gradual accumulation of point mutations. I explain the significance of moving large sequences in the next item.

7. Randomness versus functionality of inherited variations. This is perhaps the biggest question of all. How does functionality, and hence teleology, arise in random processes? My short answer is that viewed from the level of molecules we may never see it.

My argument develops in just a few stages.

(a) *Randomness and order at different levels in physics.* At a molecular level, a gas or a liquid shows random movement as the molecules interact with each other's motions. In an enclosed elastic container the global variables, pressure, volume, and temperature obey predictable laws. In our most basic science, physics, therefore, stochasticity at a low level does not entail stochasticity at a higher level. If we were to visualize a water molecule as the size of a billiard ball, the edge of the biological cell containing it would be one kilometer away. Imagine a billiard table one kilometer across and millions of billiard balls interacting. From their individual behavior we would have no idea where the constraint lies. Without that information we would not be able to work out what is going on. This is the basis of Erwin Schrödinger's argument in *What is Life?* (1943) when he said that physics is "order from disorder."

The great Dutch-Jewish philosopher Baruch (Benedict) Spinoza (1632–1677) recognized this way back in 1665 when he wrote:[28]

> Let us imagine, with your permission, a little worm, living in the blood, able to distinguish by sight the particles of blood, lymph, etc., and to reflect on the manner in which each particle, on meeting with another particle, either is repulsed, or communicates a portion of its own motion. This little worm would live in the blood, in the same way as we live in a part of the universe, and would consider each particle of blood, not as a part, but as a whole. He would be unable to determine, how all the parts are modified by the general nature of blood, and are compelled by it to adapt themselves, so as to stand in a fixed relation to one another.

Demonstrating random behavior of molecules like DNA cannot therefore exclude ordered or functional behavior at a higher level.

(b) *Organisms can use stochasticity to evolve new functions.* The next stage in my argument is that organisms have been demonstrated to use stochasticity in effective functional ways. The best example is the immune system. The germ line has only a finite amount of DNA. In order to react to many different antigens, lymphocytes "evolve" quickly to generate extensive antigen-binding variability. There can be as many as 1012 different antibody specificities in the mammalian

---

[28] B. Spinoza, "Letter to Henry Oldenburg, secretary of the Royal Society, 20 November 1665." (In *idem, The Letters,* tr. S. Shirley. Indianapolis: Hackett Publishing Co., 1995; Letter #32, pp. 192–198. The text cites an earlier translation by R.H.M. Elwes.—Ed.)

immune system, and the detailed mechanisms for achieving this have been known for many years. The mechanism is directed, because the binding of the antigen to the antibody itself activates the proliferation process. The antigen activates special lymphocytes (cells in the bloodstream) called B-cells, which evolve rapidly to generate a huge range of antigen-binding variability. Targeted speeding-up of change is therefore one mechanism by which functional change can occur. That is true even if the individual changes at that location are random. The functionality lies in the targeting of the location. That targeting is not random. It would not be functional if it was not targeted.

(c) *Natural genetic engineering has occurred during evolution.* From the Nature paper of 2001 announcing the draft sequence of the human genome,[29] two classes of proteins were shown to have evolved through transposition of complete functional domains. The details can be found in James A. Shapiro's book, *Evolution: A View from the 21st Century* (FT Press, 2011). To appreciate the full significance of these mechanisms by which whole domains can be moved around in the genome, imagine two children playing with a construction kit like Lego. To one child we give a pile of the original simple Lego bricks. To the other we also include many preformed shapes. It is obvious that when asked to make any construction that requires the preformed shapes, the second child will succeed much faster than the first. In the same way, evolution is much more likely to generate successful novel organisms if it can "play" with preformed DNA domains. Existing functionality is transferred into forming new combinations.

James Barham

We understand that your credentials to be participating in these debates at all have indeed been challenged. Without necessarily naming names, could you give us an example of such a case? What exactly did your critics say? How did you respond?

---

[29] E.S. Lander, et al., "Initial Sequencing and Analysis of the Human Genome," *Nature*, 2001, **409**: 860–921.

Denis Noble

These examples are almost hilarious, or they would be if they had not been seriously penned by chairholders in major universities:

- *Here we go again: someone arguing that DARWIN WAS RONG [sic]!*

    ‣ REPLY: I argue that Darwin was (largely) right! He didn't even read the article.

- *His most moronic claim by far is the one on mutations not being random . . . What we mean by "random" is that mutations occur regardless of whether they would be good for the organism.*

    ‣ REPLY: The potentially functional nature of some of the variations is the central theme of the articles and lectures. It is hard to miss that theme if one reads the article even cursorily. A much fuller reply is what I have written above about randomness.

- *Cells are transitory, and DNA is not.*

    ‣ REPLY: This is a common mantra, copied from Dawkins's *The Selfish Gene*. It is linguistically incoherent and factually incorrect.[30]

James Barham

As alluded to above, the most striking thing about living things, in comparison with non-living systems, is their teleological organization—meaning the way in which all of the local physical and

---

[30] D. Noble, "Neo-Darwinism, the Modern Synthesis, and Selfish Genes: Are They of Use in Physiology?," *Journal of Physiology,* 2011, **589**: 1007–1015; see, also, D. Noble, "Immortal Genes?" See Music of Life website. [http://musicoflife.website/Answers-immo rtal%20genes.html.]

chemical interactions cohere in such a way as to maintain the overall system in existence. Moreover, it is virtually impossible to speak of living beings for any length of time without using teleological and normative language—words like "goal," "purpose," "meaning," "correct/incorrect," "success/failure," etc.

Why do you think this is? We would like you to answer this question in two stages.

First, do you see teleology as objectively there in organisms themselves, or as a kind of illusion projected onto organisms by us? Doesn't cybernetic theory, together with evolutionary theory (as delineated by Harvard ornithologist and philosopher Ernst Mayr (1904–2005) and others), obviate the need for us to take the apparent teleology in living things at face value?[31]

Denis Noble

I also originally held that view. It was a debate with the Canadian philosopher Charles Taylor in 1967 that began the process by which I came to a very different view. In the first article,[32] I demonstrated what Ernst Mayr and others argue, which is that for every high-level description there must exist a valid low-level description. Taylor replied[33] that that may be true in any given case but that it would not explain what is happening if one takes a set of cases. They may be ordered only at the high level. I further replied[34] that this move makes the issue one of explanation, i.e., conceptual rather than strictly empirical.

I now go much further. My work on heart rhythm taught me that the rhythm simply doesn't exist at the molecular level. If I placed all the molecular components in a nutrient solution, but without being constrained by a living cell, the rhythm would not exist. By the usual

---

[31] E. Mayr, "The Multiple Meanings of Teleological," in *idem, Toward a New Philosophy of Biology.* Cambridge, MA: Harvard University Press, 1988; pp. 38–66.

[32] D. Noble, "Charles Taylor on Teleological Explanation," *Analysis,* 1967, **27**: 96–103.

[33] C. Taylor, "Teleological Explanation: A Reply to Denis Noble," *Analysis,* 1967, **27**: 141–143.

[34] D. Noble, "The Conceptualist View of Teleology," *Analysis,* 1967, **28**: 62–63.

ontological criteria, the rhythm doesn't exist at a molecular level but does exist at a cellular level.

James Barham

Second, if we must understand teleology in biology as objectively real, how can we do so without bringing in unwanted theological or similar baggage?

Denis Noble

I will give a brief answer here. I will explain the concepts of biological relativity and the relativity of epistemology later. The brief answer is that explaining purpose in organisms can be complete at any level, without having to go further to higher levels. The rhythm of the heart is explained at a cellular level. Its function is explained at the level of the cardiovascular system. That doesn't mean that there could not be a theological explanation. It does mean that the theological explanation is not necessary.

James Barham

One way of elucidating the difference between the objective and the mainstream views of teleology is by reference to the notion of a "machine."

A machine is a goal-directed system (it has a "function"), in which the goal state of the system is determined by an outside observer/agent. In such a system, the physico-chemical properties of the system have been carefully assembled by the outside observer/agent in order to bring about the goal state. *There is no inherent, internal, or intrinsic* tendency of the component parts and processes of the system, considered just in themselves, to produce the goal state as a distinguished system state. The intentionality and the outside intervention of the observer/agent are constitutive of what it is for something to be a machine.

A living organism is very different. Its goal state is its own continued existence, and all of its component parts spontaneously behave in such a way as to contribute to the realization of that distinguished system state (those individual contributions are then sub-goal-states). The functionality—that is, the teleology—of the system as a whole arises as a result of *purely inherent, internal, or intrinsic* processes.

Do you agree with this characterization of the essential difference between machines and organisms? That is, do you agree with the proposition that organisms are *not* machines—that machines and organisms belong to fundamentally different ontological categories?

Denis Noble

I don't think they arise as a result of "purely inherent, internal, or intrinsic processes."

They arise because organisms are open systems interacting extensively with their environment, including the behavior of other organisms. Just as such interactions can canalize the behavior of any complex but flexible system (e.g., learning machines) towards effective solutions, so the interaction with other organisms creates new solutions. The social interactions of organisms are critical to their evolution.

James Barham

What is your definition of "neo-Darwinism" (or "Modern Synthesis")? Isn't it enough to talk of "extending" it? Why do you speak of its needing to be "replaced"?

Denis Noble

The main ideas in the original formulation of neo-Darwinism are the following:

> **1. All changes in the genetic material are random,** as Weismann
> first proposed. It is important to note that what is meant is that genetic
> change is random with respect to function. This is the "blind chance"
> part of the theory with no room for teleology.

2. The germ line cells are completely isolated from the rest of the organism. Dawkins encapsulated this view in *The Selfish Gene*: "Sealed off from the outside world."

There are also some negative statements. Most important is the exclusion of Lamarckian forms of inheritance, which is implied by 2. Some modern defenders of neo-Darwinism claim to accommodate the Lamarckian forms that have now been discovered, but this seems to me to be contrary to common sense.[35] The central assumption of neo-Darwinism is that the inheritance of acquired characteristics is impossible. But if inheritable variations are not always random with respect to physiological function, then it seems to me to be more honest to say so.

I am very sympathetic to the extension idea in science. But when neo-Darwinism goes so far as to accept the inheritance of acquired characteristics and of non-random functional variations, I think we are talking more about a replacement. Note, however, that even replaced theories still have ranges of validity. Newtonian mechanics was replaced by quantum mechanics and relativity theory, but we still use Newton's equations for many ranges of applications.

I am not saying that all the equations of, e.g., population genetics, suddenly become invalid. In their range of application they still work. I am saying that neo-Darwinism is incomplete.

James Barham

You have written that "Darwin was not a neo-Darwinist."[36] We assume you mean by this something more than the obvious fact that Darwin was unacquainted with modern genetics. What exactly did you mean?

---

[35] D. Noble, "Evolution Beyond Neo-Darwinism: A New Conceptual Framework," *Journal of Experimental Biology*, 2015, **218**: 7–13; p. 7.

[36] *Ibid.*

Denis Noble

If neo-Darwinism excludes the inheritance of acquired characteristics, then Darwin was clearly not a neo-Darwinist. There are many places in *On the Origin of Species* (1859) where Darwin assumes such mechanisms and in a later book, *The Variation of Animals and Plants under Domestication* (1868), he even spelt out his theory of "gemmules" to explain them. The theory of gemmules is that chemicals pass through the blood stream to influence the germ line. We now know that RNAs do precisely that.

James Barham

What about the role of genes in life? Nowadays, many people are saying that the word "gene" has acquired so many senses as to be almost meaningless.[37] What is your own working definition of a "gene"?

Denis Noble

I agree that we now have so many definitions of "gene" that some even question the utility of the concept. Wilhelm Johannsen (1857–1927) introduced the word (*"Gen,"* in German) in 1909 as essentially a Mendelian factor. Anything (*"ein etwas"*) that determines the phenotype. That was an interpretation of Mendel's discoveries that meant that a "gene" necessarily exists when a phenotype trait obeys Mendel's laws. Of course, it was assumed that it was to be found somewhere in the organism.

The modern definition is a particular DNA sequence with start and stop codons. These definitions have very different consequences

---

[37] See, e.g., P.J. Beurton, et al., eds., *The Concept of the Gene in Development and Evolution: Historical and Epistemological Perspectives.* Cambridge: Cambridge University Press, 2001; L. Moss, *What Genes Can't Do.* Cambridge, MA: MIT Press, 2003; and P. Griffiths and K. Stotz, *Genes and Philosophy: An Introduction.* Cambridge: Cambridge University Press, 2013.

for evolutionary biology.[38] Many evolutionary biologists slip easily between the two. That doesn't work. To take just one example, I may be selfish at a phenotype level, but my selfishness would depend on my genes (and their products, proteins and RNAs) being cooperative. I try to distinguish clearly between the various definitions to avoid the pitfalls of confusing them.

James Barham

How has your work in physiology led you to challenge the "geno-centrism" of mainstream biology and evolutionary theory? What is the best way to think about the relationship between genes and organisms?

Denis Noble

Yes, that is how I got back into working on evolutionary theory. The sequencing of the genomes of many species has greatly illuminated our understanding of evolutionary biology. But it has not led to much success in enabling the development of new therapies. Even the leaders of the Project admit that the outcome has been disappointing.[39] In fact, it has been disastrous. The output of the pharmaceutical industry has declined to become pitifully small while the investment has ballooned enormously.[40]

The reason why it all went wrong is that genes do not fit what was expected of them. Very few ailments indeed depend on a single gene. Most are complex interactions involving many components in networks that extend in the body well beyond the genome. Moreover, the correlations of illness with lifestyle and family history are far

---

[38] D. Noble, "Central Tenets of neo-Darwinism Broken. Response to 'Neo-Darwinism is Just Fine,'" *Journal of Experimental Biology,* 2015, **218**: 2659–2659; and D. Noble, "Evolution Beyond neo-Darwinism," *Journal of Experimental Biology,* 2015, **218**: 7–13.

[39] Editorial, "The Human Genome at Ten," *Nature,* 2010, **464**: 649–650; and M.J. Joyner and F.G. Prendergast, "Chasing Mendel: Five Questions for Personalized Medicine," *Journal of Physiology,* 2014, **592**: 2381–2388.

[40] J.W. Scannell, A. Blanckley, H. Boldon, and B. Warrington, "Diagnosing the Decline in Pharmaceutical R&D Efficiency," *Nature Reviews Drug Discovery,* 2012, **11**: 191–200.

stronger than any correlations with the genome. People thinking of buying into commercial organizations that promise the advantages of genome sequencing should reflect on that fact. You can get better prediction at far lower cost.

James Barham

We have all read about the recent "epigenetic" revolution in our understanding of the workings of gene expression and inheritance. Another similar development is the flowering of the field known as "evolutionary developmental systems theory" ("evo-devo," for short), in which evolutionary changes are derived from changes in the trajectories of the developing embryo.

These developments are widely seen as "extending" the neo-Darwinian synthesis, rather than rivalling it or overthrowing it.

What is your view of epigenetics and evo-devo? How do your own ideas relate to these new bodies of research? Just how radical do you think they really are?

Denis Noble

It depends on the trans-generational inheritance of epigenetic changes. Since that underpins new forms of Lamarckian evolution (function influencing variation), it is clearly incompatible with neo-Darwinism and is therefore a radical change. There are neo-Darwinists who claim that it is compatible,[41] but I think that is going too far from the original definition.[42]

James Barham

Would you say that your criticism of neo-Darwinism (the Modern Synthesis) is primarily conceptual, as in for example your recent

---

[41] C.A. Williams, "Neo-Darwinism is Just Fine," *Journal of Experimental Biology*, 2015, **218**: 2658–2659.

[42] D. Noble, "Central Tenets of neo-Darwinism Broken. Response to 'Neo-Darwinism is Just Fine,'" *Journal of Experimental Biology*, 2015, **218**: 2659–2659.

paper in the *Journal of Experimental Biology*?[43] Or would you say it is primarily empirical? And how do those two things relate to each other?

Denis Noble

They necessarily go together. As an example, if the Weismann Barrier were really shown to be watertight, it wouldn't even make sense to talk about the genome as "an organ of the cell" responding to the environment, to quote Barbara McClintock.[44] One of the problems with the hardening of the Modern Synthesis during the mid-twentieth century is that it made it difficult both to think that certain experiments (e.g., on Lamarckian forms of inheritance) were worth doing and, even if one thought they were, there wouldn't be funding to do so. Many of the problems with neo-Darwinism arise from the misuse of language and the influence that has on our thought patterns. That is why I wrote the *Journal of Experimental Biology* paper. But many of my other papers deal with the empirical evidence for changing our ideas on evolutionary biology.

James Barham

You recently co-edited a special issue of *Interface Focus*,[45] a journal dedicated to exploring the "interface" between physics and biology (according to one dictionary definition, an "interface" is "a point where two systems, subjects, organizations, etc., meet and interact").

One of your co-editors was George Ellis, a well-known physicist specializing in cosmology and complexity theory. You obviously represented the other—the biology—side of the "interface." But there was also a third co-editor, namely Timothy O'Connor, a distin-

---

[43] D. Noble, "Evolution Beyond Neo-Darwinism: A New Conceptual Framework," *Journal of Experimental Biology*, 2015, **218**: 7–13.

[44] B. McClintock, "The Significance of Responses of the Genome to Challenge," *Science,* 1984, **226**: 792–801.

[45] G.F.R. Ellis, D. Noble, and T. O'Connor, eds., "Top-Down Causation," special issue of *Interface Focus,* 6 February 2012, **2**(1): 1–140.

guished analytical philosopher specializing in philosophy of mind, the philosophy of action, and the philosophy of religion.

We are very curious about how this collaboration went. Do you find it fruitful to interact with philosophers? What do you think they have to bring to the table in important scientific disputes like the one surrounding the adequacy of the neo-Darwinian explanatory framework? Would you recommend more collaborations of this sort between scientists and philosophers?

Denis Noble

I have been interacting with professional philosophers ever since I gatecrashed Stuart Hampshire's graduate philosophy class at University College London 57 years ago. I first published in a professional philosophy journal 49 years ago. The professional philosophers I have seriously interacted with since include R.M. Hare, Charles Taylor, Bernard Williams, Alan Montefiore, Anthony Kenny, Peter Hacker, Bryan Magee, Daniel Dennett, and Jos de Mul. There are many more. One highly respected British philosopher acknowledges me as a philosopher of science, in addition to being a practicing scientist, while another recently completely mistook me for being only a professional philosopher. It seems to run in the family. Like my brother, Raymond Noble, with whom I share many views on evolution, I am an academic chameleon.

Such interaction used to be much more common. The world of science today makes it very difficult. This is unfortunate. As the great French physicist and polymath Henri Poincaré (1854–1912) remarked a century ago, those who claim they are not philosophers make the worst conceptual errors.[46] They don't even see the conceptual holes into which they fall.

---

[46] H. Poincaré, *La science et l'hypothèse*. Paris: Flammarion, 1902. (Translated as *Science and Hypothesis*, numerous editions.—Ed.)

James Barham

Why do you think the interpretation of evolutionary biology raises such passion, and even anger?

Denis Noble

Two opposing sides became entrenched into dogmatism: the creationists on one side and the neo-Darwinists on the other. It is as simple as that, a kind of war of religion, since dogmatic science shouldn't exist. When it does it becomes a faith rather than a science. Faith wars almost invariably engender anger and cruelty. Remember the cruelty of the Crusades.

James Barham

We see from your website that you have been traveling the world (including China) giving lectures about your anti-reductionist view of life. We are curious to know what sort of reception you've been having.

Are you finding that your audiences are more open to your message than they used to be? What about the Chinese, in particular —how willing are they to entertain new ideas about the nature of life?

Denis Noble

In many parts of the world the clash between creationism and dogmatic neo-Darwinism doesn't have anything like as much impact. Many religious traditions find evolutionary biology to be no challenge to them. Buddhism in Tibet, Japan, China, Korea, Thailand, Burma, Vietnam, and Sri Lanka, as well as the original religions, such as Shinto, simply don't have a problem. The only times I have encountered problems with discussing ideas of evolution in those parts of the world has been where strongly fundamentalist forms of western religions have penetrated South and East Asia.

I have argued elsewhere[47] that part of the reason is that these are religions in which ritual and practice are much more important than belief systems. It is for that reason that many Buddhists, for example, argue that their tradition is not a religion. I have also found that in my interactions with the Dalai Lama and other leading Buddhists they are wide open to the findings of empirical science.

James Barham

Recently, you have been involved with a number of like-minded scientists in launching a website devoted to spreading anti-reductionist ideas about the nature of life. It is called the Third Way of Evolution.[48]

Could you tell us a bit about how this project came into being? What was your role in it? Besides you, who else was involved in founding the website? What does your group hope to achieve?

Denis Noble

I have interacted with many evolutionary biologists, representing both old and new views, and over many decades. A recurring theme amongst those who question aspects of neo-Darwinism, whether proposing extension or replacement, is that they often experience frankly insulting remarks from some of the more dogmatic neo-Darwinists (I emphasise that this is not true of all neo-Darwinists by any means) and that they have much greater difficulty getting articles published.

When James Shapiro and Raju Pookottil told me of their idea to launch a Third Way of Evolution website, I readily agreed. I therefore became one of the three founders. When I meet with those (more than 50 now) who have joined it, I find that they are very pleased that we took this initiative.

---

[47] D. Noble, "A Systems Biological Interpretation of the Concept of No-Self (*anāt-man*)," in Venerable Chuan Sheng, ed., *Exploring Buddhism and Science*. Singapore: Buddhist College of Singapore/Kong Meng San Phor Kark See Monastery, 2015; pp. 234–260.

[48] [https://thethirdwayofevolution.com/.]

It also establishes a very important point. Criticism of neo-Darwinism should not be taken to mean support for creationism or intelligent design. We have put a marker down. It is now much easier to dissent, without being accused of supporting creationism or intelligent design. That alone makes the initiative worthwhile.

James Barham

On the homepage of the Third Way website is the following statement:

"It has come to our attention that THE THIRD WAY web site is wrongly being referenced by proponents of Intelligent Design and creationist ideas as support for their arguments. We intend to make it clear that the website and scientists listed on the website do not support or subscribe to any proposals that resort to inscrutable divine forces or supernatural intervention, whether they are called Creationism, Intelligent Design, or anything else."

I would like to play Devil's Advocate for a moment and ask you to support this claim further. After all, in the minds of a great many people, "teleology," "intelligence," and "design," as they are manifested in biology, are pretty nearly three names for the same thing.

Therefore, could you please expand on the difference, as you see it, between your approach to teleology in biology and that of the Intelligent Design people?

Denis Noble

This naturally follows on from the previous answer. My view is that teleology is alive and well in biology because organisms clearly have goals, including feeding and reproduction, and the natural intelligence to realize them. Moreover, it is just as easy for biologists to test theories about goals in organisms as it is for engineers to test such behavior on man-made systems.

The work of English physiologist William Harvey (1578–1657) on the circulation of the blood made it possible to understand the purpose of the heart and circulation and led to the search for capillaries.

Good teleological theories have precisely such valuable outcomes in experimental science.

The difference from creationism and intelligent design is that we think that evolution has produced organisms with purpose.

James Barham

You have spoken of the "beguiling" nature of the mainstream, reductionistic interpretation of evolution that we find in neo-Darwinism.[49] You have also said that "it is almost impossible to stand outside it."[50] We have two questions for you in this connection.

First, why do you think the neo-Darwinian view is so beguiling? Why are so many people so attached to the notion that they are "nothing but" animals, and that living things are "nothing but" the atoms and molecules that compose them?

We would have thought this view of things would be depressing, but we have the impression that many people would rather undergo any degree of intellectual contortion than relinquish their cherished adherence to the mechanistic and reductionist neo-Darwinian worldview.

Denis Noble

Neo-Darwinism is beguiling and convincing because it was developed by some of the best biological minds of the twentieth century and has been popularized by extremely clever writers who, themselves, felt that they were on a mission to convince the world of the insights of their science. They succeeded. Their theories thoroughly permeate the social sciences (sociobiology, and game theory models in economics and sociology), politics (neo-conservatism), literature (which freely uses their colorful metaphors), politics (the Cold War divide was partly buttressed by polarization over Lysenkoism—in my view a very bad version of Lamarckism), and so it goes on. I

---

[49] D. Noble, "Evolution Beyond Neo-Darwinism: A New Conceptual Framework," *Journal of Experimental Biology*, 2015, 218: 7–13; p. 7.

[50] *Ibid.*

simply don't know of an area of human endeavor that has not been influenced.

Is it depressing? I quote the ending of a review of The Music of Life[51] by the distinguished Dutch philosopher, Jos de Mul:

> Noble offers a powerful antidote to the nihilism of Dawkins. Although Dawkins writes on the last pages of The Selfish Gene that man is the only creature to rebel against the selfish genes, how [would that] be possible in the light of the reductionist determinism which permeates the preceding two hundred pages of his book [and] remains completely unresolved[?]

That reassuring incantation is not always received by Dawkins's readers. I had to think about it when I read the interview that the well-known Dutch author Joost Zwagerman gave to [the Dutch newspaper] HP/De Tijd four days before his self-chosen death. Referring to a statement by Nietzsche, he says that the thought of suicide for a long time gave him consolation during bad times in his life. But that comforting character completely disappeared when his father undertook an attempt to take his own life. From that moment his life was dominated by the fear that he and his children and future grandchildren would be genetically predisposed to commit suicide.

Of course, I do not claim that the neo-Darwinian view of man alone drove Zwagerman to suicide. The failure of his marriage, the incurable, very discomfiting and painful ankylosing spondylitis, and recurrent depression will undoubtedly have also played an important role. Again, it is always a combination of elements in life. But the idea of genetic predestination found in books like The Selfish Gene seems to me very likely to have played a role.

I don't think I need add anything to Jos de Mul's remarks.

James Barham
Second, what does it take to break the charm?

---

[51] [http://www.musicoflife.website/pdfs/De%20Mul%202016%20Noble%20versus%20Dawkins.pdf.]

This issue is of enormous public significance, especially for the way we view moral and legal responsibility, but also in medicine, in politics, and in other domains of public life.

What is the best way to help the public to see that the mainstream reductionist view of life, however beguiling, is not nearly so well supported scientifically as they think?

Denis Noble

My belief is that those of us who think differently need to try at least to emulate the literary skills of the popularizing neo-Darwinists. That is why I wrote *The Music of Life*, and why I recently wrote a second book that goes much further than the first one: *Dance to the Tune of Life: Biological Relativity*.

James Barham

One of the terms you use to describe your own alternative position is "biological relativity." Could you explain to us, briefly and in non-technical terms if at all possible, what you mean by this phrase?

Denis Noble

This is one of the themes of the new book referred to above. Briefly, the central problem with neo-Darwinism, as with reductionist biology in general, is that it makes some (usually hidden) metaphysical assumptions. To quote James D. Watson:[52]

> There are only molecules—everything else is sociology.

To many people, not only scientists, this now seems almost obvious.

I was not myself immune from this feeling. When I first read Dawkins's *The Selfish Gene* I thought, "Wow, are we really determined by a set of molecular sequences that work inside us like viruses?" It feels almost creepy until you get to the last chapter, when Richard makes some important and revealing disclaimers. You get

---

[52] Attributed by V.S. Ramachandran, "Guest Editorial: The Astonishing Francis Crick," *Perception*, 2004, **33**: 1151–1154; p. 1152.

drawn in by the colorful metaphors, greatly helped by the ease with which the reductionist story can be told. Complexity is much more difficult to expound to a general audience.

The more I thought about this, the more I realized that it would take a root-and-branch approach to counter it. The central idea ("there are only molecules") is what needs challenging. That is clearly not true. Why not say "there are only strings," or whatever physicists now identify as the most fundamental (note the force of this word) entities? But more importantly, why should we think that the universe cares about any particular level?

The Theory of Biological Relativity answers that question by showing that there is no justifiable basis for privileging the molecular level. By doing so, neo-Darwinism automatically excludes teleology, because there is no form of life at that level. DNA alone is dead. The Theory of Biological Relativity also challenges a metaphysical assumption in the gene-centric approach. This is that, because there can be no purpose at the molecular level, there can be no purpose anywhere. Blind chance is seen here as the "real" underlying nature of the universe. That view is supported by the feeling that, at the lowest scale, the fundamental particles cannot conceivably have goals or intentions. But there is also a further assumption, which is usually unspoken. This is that if the elements cannot have goals, then nothing else in the universe can do so. There is no room in this interpretation for purpose.

Purposive behavior arises precisely because evolution enables it to do so. The nihilist version of neo-Darwinism therefore denies evolution's own major achievement, the creation of purpose!!

And, by the way, instead it attributes (metaphorically?) the missing purposiveness to genes. They alone are allowed to have a goal: to be selfish. The rest is a throwaway.

James Barham

Nonlinear dynamical models of biofunctions have been dismissed by some observers as "phenomenological," meaning they may simulate aspects of the behavior of biological systems more or less well, but they do not make contact with the underlying causal processes—i.e., the chemistry and physics—and thus ultimately have only limited explanatory value.

Others, like Stuart A. Kauffman,[53] Gerald H. Pollack,[54] the late Mae-Wan Ho,[55] Alexei Kurakin,[56] and several others have counseled more radical approaches based on "collective phenomena" arising out of the condensed-matter-physics properties of "the living state of matter."

What do you make of such proposals? Do you see them as a promising avenue of research? Or is all of this a bridge too far, in your opinion?

Denis Noble

I think that challenging the idea of privileging the molecular level is enough. Everything else in my work flows from that.

James Barham

In closing, we would like you to tell us—in bulleted list format, if you like—what you consider to be the five strongest arguments in support of your view that "neo-Darwinian is not enough," as well as the five weakest arguments that supporters of neo-Darwinism commonly advance.

Denis Noble

Neo-Darwinism is not enough because:

---

[53] S.A. Kauffman, *Humanity in a Creative Universe*. Oxford: Oxford University Press, 2016.

[54] G.H. Pollack, *Cells, Gels, and the Engines of Life: A New, Unifying Approach to Cell Function*. Seattle: Ebner & Sons, 2001.

[55] M.-W. Ho, *The Rainbow and the Worm: The Physics of Life*, 3rd ed. Singapore: World Scientific, 2008.

[56] A. Kurakin, "Scale-free Flow of Life: On the Biology, Economics, and Physics of the Cell," *Theoretical Biology and Medical Modelling*, 2009, **6**: 6; doi: 10.1186/1742-4682-6-6.

1. The gene-centric view has failed in one of its major claims, i.e., that it would result, through sequencing genes, in curing the major diseases that plague humanity.

2. It doesn't have a sound metaphysical basis. There is no justification for privileging any one level in biological systems. No one has ever produced such a justification.

3. It has had profoundly damaging (even if not intended) consequences in sociology, economics, politics, and many other areas of the humanities and social sciences.

4. It has had to gyrate in a contorted way to accommodate one new finding after another. The final straw for me was a supporter of neo-Darwinism purporting to accept the inheritance of acquired characteristics. This is like eating your own tail.

5. Its claim to parsimony. Nature simply isn't parsimonious.

I believe that those five points answer both questions. But let's try to formulate the other five:

1. The claim that the Weismann Barrier and the Central Dogma have settled the question whether Lamarckism is possible. But Weismann's experiments were not a test for Lamarckism and the Central Dogma does not counter the fact that the organism controls the genome.

2. The claim that epigenetic inheritance always dies out after a generation or two. There are clear examples where it doesn't, and, in any case, no one supposes that an evolutionary change initiated by epigenetic effects would be the consequence of a single-generation exposure to the changed environment. Multiple-generation exposures can be assimilated into the genome.

3. The claim that genetic change is always random with respect to function. It is almost certain that it would be, since randomness at the molecular level is what you would expect even if functionality exists at other levels.

4. Neo-Darwinism is obvious and necessarily true (often advanced by Dawkins, as in the debate with Margulis). If it were, it would become a tautology and not open to experimental verification. Not much good as a scientific theory.

5. It was formulated by some of the greatest scientists of the twentieth century, so it must be right. *Nullius in Verba!*[57]

James Barham

Finally, what are your plans? We take it you are still actively engaged in research: What can share with us about your current projects?

Denis Noble

As to active research, I am trying to lead the way back into investigating natural products in healthcare. The failure of the Human Genome Project to produce more than a handful of successful medications compels us to look elsewhere.

Remember, too, that about half of the drugs in Western pharmacopeias came from natural products. The 2015 Nobel Prize for Physiology and Medicine was awarded for this kind of work—to Tu Youyou, in China.[58]

James Barham

On behalf of ourselves and our readers, we would like to thank you for sharing your time and your thoughts with us.

We also look forward eagerly to the Dialogue on Evolution between you and David Sloan Wilson.

---

[57] The motto of the Royal Society of London, founded in 1660, meaning "On the word of no one"—i.e., "Don't take anyone's word for it!"—Ed.

[58] Y. Tu, Nobel Prize Lecture, 2015. [https://www.nobelprize.org/prizes/medicine/2015/tu/lecture/.]

3. David Sloan Wilson's Major Statement: The Neo-Darwinian Revolution Is Far from Complete

I thank AcademicInfluence.com for sponsoring this dialogue and pairing me with Denis Noble. In some ways I am an odd choice to argue for the position that neo-Darwinism is enough, since I am a well-known critic of selfish gene orthodoxy and a participant in the first workshop to use the title "extended evolutionary synthesis."[1] Nevertheless, I am happy to argue for this position as long as it is appropriately framed.

My programmatic activities during the last 13 years also qualify me for the job: First, as director of EvoS[2] (standing for "Evolutionary Studies," and pronounced as one word), which teaches evolution across the curriculum at Binghamton University, and then as co-founder and President of the Evolution Institute (EI),[3] which applies evolutionary theory to public policy issues of all sorts. Through a number of communication outlets managed or supported by the EI (This View of Life, Social Evolution Forum,[4] Evonomics.com, and

---

[1] M. Pigliucci and G.B. Müller, eds., *Evolution—The Extended Synthesis*. Cambridge, MA: MIT Press, 2010.

[2] [https://www.binghamton.edu/evos/.]

[3] [https://evolution-institute.org/.]

[4] [https://evolution-institute.org/blog/social-evolution-forum-2012-in-review/.]

PROSOCIAL Magazine[5]), I have taken on the role of a newspaper reporter, covering the beat of evolutionary science. I will draw upon this experience, in addition to my own research, in what follows.

3.1. Defining Neo-Darwinism

One of the most elegant distillations of conventional evolutionary theory was by the pioneering Dutch ethologist Niko Tinbergen, in an article titled "On aims and methods of ethology" (1963).[6] At the time, it was not widely accepted that a behavioral trait (such as aggression) can evolve in the same way as a physiological or anatomical trait (such as a deer's antlers). In the process of making this case, Tinbergen observed that four questions need to be addressed for all products of evolution, concerning their function, mechanism, development, and phylogeny.

The function question concerns why the trait exists, compared to many other traits that could exist, often because of the winnowing action of natural selection. The mechanism question concerns how the trait exists in a physical sense. The development question concerns how the trait comes into being during the lifetime of the organism. The phylogeny question concerns how the trait came into being over multiple generations, since evolution is a historical process.

Tinbergen's case that behaviors evolve much like other traits has become so thoroughly accepted that no one questions it anymore. He shared the 1973 Nobel Prize in Physiology or Medicine with Konrad Lorenz and Carl von Frisch for their pioneering work in ethology.[7] Tinbergen's four questions, however, are still widely cited as a compact description of a fully rounded evolutionary perspective. I will take them as my definition of neo-Darwinism.

---

[5] [https://web.archive.org/web/20170716204637/http://magazine.prosocialgroups.org /.]

[6] N. Tinbergen, "On Aims and Methods of Ethology," *Zeitschrift für Tierpsychologie*, 1963, **20**: 410–433.

[7] The scientific study of animal behavior from an ecological and evolutionary perspective—Ed.

Some might disagree. For example, it is often said that the so-called Modern Synthesis left out development. In this regard, it is notable that Harvard ornithologist Ernst Mayr, one of the main names associated with the Modern Synthesis, made a two-fold distinction between ultimate and proximate causation,[8] rather than Tinbergen's four-fold distinction.

Roughly, Tinbergen's function and history questions map onto Mayr's concept of ultimate causation, while the mechanism and development questions map onto the concept of proximate causation. So Mayr's terminology did push development into the shadows by failing to distinguish between the adult trait and how it comes to exist during the lifetime of the organism. Nevertheless, there is no reason to privilege Mayr over Tinbergen in their characterization of evolutionary science during the 1960s.

Having clearly stated my definition, here is the position that I will defend in this essay: Tinbergen's four questions are enough for the future progress of evolutionary theory. In fact, to claim otherwise is a distraction.

3.2. The Neo-Darwinian Revolution is Not Yet Complete!

It is shocking, when one pauses to think about it, how many fields of inquiry do NOT employ a fully rounded four-question approach. In this sense, the neo-Darwinian revolution is very far from complete. An example will illustrate the enormity of the problem.

James Coan is a highly regarded clinical psychologist and neuro-scientist at the University of Virginia's Department of Psychology. In many respects, he is at the cutting edge of his field, employing brain imaging techniques to investigate the nature of problems such as post-traumatic stress disorder (PTSD) in war veterans. Yet, difficulties interpreting a line of experiments and a suggestion by a colleague

---

[8] E. Mayr, "Cause and Effect in Biology," *Science*, 1961, **134**(3489): 1501–1506.

to read a book titled *An Introduction to Behavioural Ecology*[9] led Coan to have what can only be called a born-again experience.

Coan read the fourth edition of this book. The first edition was published in 1981 and reflected a maturation of what Tinbergen, Lorenz, and von Frisch had started. Especially powerful was the maturation of the "function" question. More and more, animal behaviorists were asking the question: "How *would* the organism that I study behave, if it were a product of natural selection?"

Asking this question allowed mathematics to be employed in a way that was new for the study of animal behavior, although customary in the field of economics (more on this below). For example, in a foraging organism that is adapted to maximize its rate of energy gain (E) per unit time (T), just a few additional assumptions about its foraging behavior (e.g., it can only handle one item at a time) and its foraging environment (e.g., each prey type, i, is randomly encountered and is characterized by a density, an amount of energy, e_i, and handling time, h_i) are sufficient to generate detailed predictions about its foraging behavior (e.g., it should rank each prey type according to its e/h ratio and always accept the highest-ranked prey types down to a cut-off point, below which it should ignore all prey types) that were highly novel and testable in the laboratory or field.

If the assumptions of this particular model don't fit a particular foraging organism, then models with different assumptions can be built (e.g., non-random food distributions, foraging under the risk of predation). In this fashion, a whole family of models called "optimal foraging theory" sprang up in the 1960s and 70s, which organized the study of foraging behavior as never before. As for foraging behavior, so also for every other behavior (e.g., mating, fighting, cooperation) and traits that are typically not considered behaviors (e.g., sex ratio, life history strategies, developmental strategies, strategies of the immune system).

---

[9] N.B. Davies, J.R. Krebs, and S.A. West, *An Introduction to Behavioural Ecology*, 4th ed. Chichester, UK: Wiley-Blackwell, 2012.

The function question helped to inform the other questions, and vice versa. White-tail deer and kangaroos are both mammalian herbivores, but the fact that they evolved on different continents makes them very different from each other, which influences their development and specific adaptations to their environments. To focus on brain mechanisms, suppose that I assigned you the task of studying the brains of two bird species without telling you anything about their ecologies. Unbeknownst to you, one species migrates south during the winter and is adapted to memorize the night sky as a nestling. The other remains north during the winter and is adapted to memorize the location of thousands of food items stored during the fall. How many decades would be required for you to discover the mechanistic basis of these adaptations, studying only the brains of the two species?

All of this was part of my training as a graduate student in the 1970s, but it was new to Coan in the 2010s and reading the fourth edition of *An Introduction to Behavioural Ecology* was like the scales falling from his eyes. Here is how he described the experience during an interview that I conducted with him in 2016:[10]

> **James Coan:** I think "meteoric" is the right way to describe its impact on me. By the time I had finished the first chapter I was already thinking about my own work, and indeed, thinking about psychology as a broad discipline completely differently. The book starts out introducing principles that organize behavior—that when you give them even a little bit of thought make complete sense. Principles like the management of bio-energetic resources; that if you're going to engage in a behavior as an organism, to accrue resources, you have to invest resources that you have in store. That is a very risky business, so you need a certain amount of information about the demand of the environment and your own resource cache. That entails certain principles that get built into the genome over time about keeping an excess, having a surplus, and maintaining a surplus. So you must balance your investment against ...

---

[10] [http://circleofwillispodcast.com/episode/497662f0a5eb4685/david-sloan-wilson.]

David Sloan Wilson: Tradeoffs, tradeoffs, tradeoffs.

JC: Tradeoffs everywhere. I had a kind of personal and intellectual crisis, where I thought "Holy shit! What have I been doing all this time? I've been thinking about constructs that aren't tethered to any ultimate goals or any ultimate constraining principles. In psychology, anything goes, because the thinking isn't constrained by these imperatives of biological organisms across evolution and ontogeny. Then I started going through chapter after chapter, example after example, of these principles existing not just as logical arguments but as empirical data. It was enough to almost make me cry.

Coan's academic disciplines of psychology and neuroscience were highly sophisticated in their own ways, but they had not converged upon Tinbergen's fully rounded four-question approach. In a sense, Coan was like someone who had been given the brain of a species to study without being told nearly enough about the ecology and evolutionary history of the species—*Homo sapiens,* in his case.

Once one becomes attuned to the four-question perspective, the extent of its absence becomes shocking. Consider this segment of an interview that I conducted with the Harvard population geneticist Richard C. Lewontin in 2015.[11] Lewontin was a student of the geneticist Theodosius Dobzhansky, who famously pronounced that "Nothing in biology makes sense except in the light of evolution" in an article written in 1973.[12]

Richard C. Lewontin: I was raised not as an evolutionist but as a population geneticist.

David Soan Wilson: Right.

---

[11] [https://web.archive.org/web/20170720184011/https://evolution-institute.org/articl e/the-spandrels-of-san-marco-revisited-an-interview-with-richard-c-lewontin/.]

[12] T. Dobzhansky, "Nothing in Biology Makes Sense Except in the Light of Evolution," *American Biology Teacher*, 1973, **35**: 125–129.

RCL: That's a big difference.

DSW: Why is that a big difference? Let's clarify that for me. I tend to see it as a small difference. What's the difference between being a population geneticist and an evolutionist?

RCL: A population geneticist by theoretical training has certain parameters of population change. That's become broadened by the realization that there are between-population changes, and so on, but within a population we're talking about changes in gene frequency and we have a catalog of the causes: selection, inbreeding, chance, mutation, and so on. Our job as population geneticists is to do the necessary observations of the various things that give us estimates of the strength of those different forces. Now, historically one of the most interesting—now I want to talk a little about the sociology of our science—Theodosius Dobzhansky, my professor and then-greatest living evolutionary biologist…

DSW: Mr. "Nothing in biology makes sense except in the light of evolution."

RCL: Yeah, right. He was a very bad field observer. Theodosius Dobzhansky never, in his entire life, nor any of his students, me included—I would go out in the field with him, actually—ever saw a *Drosophila* pseudoobscura in its natural habitat.

DSW: (laughs) Yeah, OK!

RCL: We didn't know where they laid their eggs. We couldn't have counted the number of eggs of different genotypes. How did we study *Drosophila* in the wild? We went out into the desert, into Death Valley, we moved into a little oasis, we went first to the grocery store, and bought rotten bananas. We mushed up the bananas with yeast till they fermented a bit, we dumped that into the paper containers, put it out in the field and the flies came to us.

DSW: Right! No naturalistic context whatsoever.

RCL: None … at … all. And to this day we do not know anything about the actual habitat of *Drosophila* pseudoobscura, although by the way, interestingly enough, in more recent years, Tim Prout actually succeeded in trapping pseudoobscura in orange groves, so we don't even know how much they hang out with cultivated fruit.

DSW: Right.

RCL: Now let me go one step further because we cannot understand the development of evolutionary biology if we don't understand questions of the sociology of academic life. If I wanted to study evolutionary forces acting on some genetic polymorphism in *Drosophila*, I would go and look for some species of *Drosophila* where I could actually look at, perturb, and work with the actual breeding sites and egg-laying sites, and pick up larvae in nature, and so on. And, in fact, there is such a group of *Drosophila*. They are the cactophilic ones. There is a group [of scientists] from Texas and other places that studies the cactophilic *Drosophila* in an ecologically sensible way of going to the rot pockets and perturbing them, getting larvae out of them, and so on. That group never acquired the prestige associated with the Dobzhansky school because—I don't know why. They were doing what one has to do. That's why, for example, I try to convince students who are entering evolutionary biology not to study animals at all but to study plants. Plants stay in one place. You can manipulate them. You can move them. Plants are much better than animals for studying things in nature. Yet, plant evolutionary biology is not, for sociological reasons that I don't understand—I could make up stories —has never had the prestige that animal work has had when it comes to population genetics.

DSW: Right. I think that [there was an] all-consuming interest in physical mechanisms as opposed to a more fully rounded approach. I place a lot of emphasis on the classic paper by Niko Tinbergen, "On aims and methods of ethology", in which he says that you have to ask four questions: Function, History, Mechanism, Development. Are you familiar with that paper?

RCL: No, I'm not. Send me a reference to it.

DSW: It's such a succinct summary of what a fully rounded approach needs to be …

Lewontin's comments reveal that Tinbergen's four-question approach is still a work in progress within the biological sciences, before we even get to more human-oriented academic disciplines.

Every major discipline such as ethology, ecology, paleontology, systematics, development, genetics, population genetics, biochemistry, and molecular biology has a separate history that is contemporaneous with the theory of evolution. Model species such as *Drosophila* were chosen during the early twentieth century because they were easy to culture in the laboratory or because their chromosomes were easy to stain and observe under the microscope. The most pressing questions were often mechanistic in nature, pushing Tinbergen's other three questions into the background. Who needs to know the nuances of the ecology of *Drosophila* when you're trying to work out the basic mechanisms of recombination, the transcription of genes into proteins, and the like?

As mechanistic knowledge increases, then the need to bring in the other questions becomes increasingly important, but integration can require decades and is impeded by a variety of intellectual and sociological factors. If I were to nominate the single most important priority for the biological sciences, it would be to get everyone on the same page with respect to Tinbergen's four questions.

A similar story can be told for medicine and the health sciences, which are obviously advanced and sophisticated in their own ways, but shockingly clueless, for the most part, about Tinbergen's four questions. That's why George C. Williams, an evolutionary biologist by training, and Randolph Nesse, a psychiatrist by training, could write a groundbreaking academic article in 1991 titled "The Dawn of Darwinian Medicine," followed by their trade book *Why We Get*

Sick in 1995.[13] Both of these works are elementary tutorials, much like the article that Tinbergen wrote in 1963, except oriented toward the topic of medicine rather than ethology. Fast-forwarding to the present, Nesse is founding director of the Center for Evolution and Medicine[14] at Arizona State University, which was established in 2015. In my 2016 interview with him,[15] he reports that a fully rounded four-question approach is only beginning to gain a toehold in the health sciences.

Let's zoom in on cancer as one topic within the health sciences. Cancer is natural selection taking place among the cells of a multicellular organism. Evolution has no foresight, so the fact that proliferating cancer cells eventually bring about their own demise with the death of the organism is immaterial. That might also be our epitaph if we succeed in causing our own extinction as a species. Since multicellular organisms that are relatively cancer-proof survive and reproduce better than those that are relatively cancer-prone, billions of years of natural selection at that level have resulted in elaborate genetic and physiological mechanisms that protect against cancer cells. Nevertheless, lower-level selection (among cells within the organism) is only suppressed by higher-level selection (among organisms), never entirely eliminated, and can erupt when evolved protective mechanisms are disrupted by modern environmental factors that were not present in our ancestral environments.

The preceding paragraph describes a fully rounded, four-question approach to the study of cancer. Once learned, employing the four-question approach is as easy as riding a bicycle. But the first scientific articles describing cancer as natural selection among cells within our bodies weren't published until the 1970s and the community of cancer researchers employing the four-question approach is a minus-

---

[13] G.C. Williams and R.M. Nesse, "The Dawn of Darwinian Medicine," *Quarterly Review of Biology*, 1991, **66**: 1–22; and R.M. Nesse and G.C. Williams, *Why We Get Sick: The New Science of Darwinian Medicine*. New York: Times Books, 1995.

[14] [https://evmed.asu.edu/.]

[15] [https://web.archive.org/web/20170705100744/https://evolution-institute.org/articl e/evolutionary-medicine-comes-of-age-an-interview-with-randolph-nesse/?source=tvol.]

cule fraction of the larger cancer research community. The dynamic is similar to what Lewontin described for genetics in the first half of the twentieth century—an obsessive focus on physical mechanisms and inattention to the other three questions.

One cancer researcher who does champion the four-question approach is Athena Aktipis, who was trained as a theoretical evolutionary biologist and studies a variety of topics centered on the evolution of cooperation in addition to cancer. In my 2016 interview with her[16] and her report on the first Evolutionary Biology and Ecology of Cancer (EBEC) Summer School[17] at the Wellcome Genome Campus[18] in the UK, she describes the same emerging paradigm that I described for the study of behavioral ecology in the 1960s and 70s, which was so new for James Coan in the areas of clinical psychology and neuroscience.

Here is an example of how elementary four-question reasoning provides new insights for the study of cancer. Since natural selection within the body requires cell divisions, the more cell divisions that take place, the higher the likelihood of cancer. This pattern is typically observed within a species. Old people get cancer more than young people (which you already knew) and tall people get cancer more than short people (which you probably didn't know—at least I didn't!). Yet, the pattern does not hold across species. The incidence of cancer in long-lived species such as elephants is not greater than in short-lived species such as mice. This is called Peto's paradox and the most likely explanation is that long-lived species have evolved better cancer-prevention mechanisms than short-lived species. Knowing this, you'd think that cancer researchers would be interested in studying long-lived species to discover new drugs, gene therapies, and the like for the treatment of cancer in humans. More generally, you'd

---

[16] [https://web.archive.org/web/20170901202515/https://evolution-institute.org/article/the-evolutionary-ecology-of-cancer-an-interview-with-athena-aktipis/.]

[17] [https://web.archive.org/web/20190331191718/https://coursesandconferences.wellcomegenomecampus.org/.]

[18] [https://web.archive.org/web/20170710223506/http://www.sanger.ac.uk/about/campus.]

think that cross-species comparisons would be an important part of cancer research, but this is not the case. The vast majority of cancer research is conducted on a very few model organisms chosen for ease of laboratory research—like the early days of *Drosophila* research described by Lewontin. Granted, it's not easy to study elephants, but the comparative study of cancer in different dog breeds provides rich opportunities for employing a fully rounded, four-question approach and this too is a new idea in cancer research.

Let's pause to take stock of the argument that I have developed so far for why neo-Darwinism is enough. If by "neo-Darwinism" we mean Tinbergen's four questions, then we can see that he was premature to suggest in 1963 that they are already being employed by biologists for the study of non-behavioral traits and merely needed to be extended to behavioral traits. A fully rounded, four-question approach also needed to be applied to many non-behavioral traits, especially in areas of the biological sciences where an excessive focus on physical mechanisms—one of the four questions—crowded out the other three questions. Moreover, this problem still exists for topic areas such as neuroscience and cancer research in the twenty-first century. No amount of sophistication in the study of proximate mechanisms can substitute for a fully rounded, four-question approach. Until this point is well understood, to say that we need to go beyond the four-question approach is a distraction.

My argument so far also highlights the need for a fine-grained understanding of the many disciplines that comprise the biological sciences. They are connected to a degree, but they also have their own histories and integration is still a work in progress. History matters —for the study of academic disciplines no less than for the study of biological species.

3.3. The Human-Related Academic Disciplines

When we move from the biological and "hard" psychological sciences (e.g., neuroscience) to the human behavioral sciences and humanities, the situation gets even worse. It is not an exaggeration to say that the study of evolution in relation to human affairs lags behind the study of evolution in the biological sciences by about a century.

As an example, *Sociobiology: The New Synthesis*, published by the evolutionary biologist Edward O. Wilson in 1975,[19] was cut from the same cloth as *An Introduction to Behavioural Ecology* and reflected the maturation of what Tinbergen, Lorenz, and von Frisch had started. Wilson's thesis was that a single theoretical framework could explain the evolution of social behaviors in all species, from microbes to humans. It was celebrated as a synthetic triumph, except for the final chapter on humans, which created an uproar, complete with charges of fascism and an infamous pitcher of water that was dumped on Wilson's head during an annual conference of the American Association for the Advancement of Science.

Over a century after the publication of Darwin's *Descent of Man* in 1871,[20] it was not permissible to study human social behavior from an evolutionary perspective. It wasn't until the 1980s that terms such as "Evolutionary Psychology," "Evolutionary Anthropology," "Evolutionary Economics," and "Literary Darwinism" began to be coined, signifying a renewed attempt to rethink the human-related academic disciplines from an evolutionary perspective–and even these had an air of scandal about them.

That's the bad news. The good news is that enormous progress has been made since then, which is reported in dozens of books and hun-

---

[19] E.O. Wilson, *Sociobiology: The New Synthesis*. Cambridge, MA: Harvard University Press, 1975.

[20] C. Darwin, *The Descent of Man, and Selection in Relation to Sex*, two vols. London: John Murray, 1871.

dreds of peer-reviewed articles without anyone raising an eyebrow. The future is already here for a sizable community of scientists and scholars who use a fully rounded, four-question approach to explore the length and breadth of humanity. However, this community is still a tiny fraction of the worldwide academic community in the human-related academic disciplines. Also, these recent developments are scarcely reflected in college-level education, as I will discuss in more detail below.

The two communication outlets that are produced by the Evolution Institute—This View of Life (TVOL) and Social Evolution Forum (SEF)—exist to catalyze the completion of the neo-Darwinian revolution. TVOL articles are written for a broad audience without being watered down and SEF blogs, target essays, and commentaries are written at a more professional level. If you visit these sites, you will see nothing more or less than the application of Tinbergen's four questions to "anything and everything" in biology, the human-related sciences, and the humanities. For example, my target essay in SEF titled "The One Culture"[21] reviews four books that are helping to unify the sciences and humanities, in contrast to the disconnect between them that the British scientist/novelist C.P. Snow called attention to in his famous 1959 Rede Lecture entitled "The Two Cultures."[22]

The Two Cultures" can't become "The One Culture" unless human culture itself can be understood from an evolutionary perspective. Progress toward this goal was impeded for most of the twentieth century, not only because of avoidance on the part of humanists, but also because evolutionary biologists became excessively gene-centric. Darwin knew nothing about genes and conceptualized evolution in terms of variation, selection, and heredity—a resemblance

---

[21] [https://web.archive.org/web/20170703090623/https://evolution-institute.org/focus-article/the-one-culture/.]

[22] C.P. Snow, *The Two Cultures*. Cambridge: Cambridge University Press, 1959; for discussion, see the Wikipedia article here. [https://en.wikipedia.org/wiki/The_Two_Cultures.]

between parents and offspring—which can be measured at the phenotypic level without any knowledge of the underlying mechanisms. Once the science of genetics was born in the early twentieth century, however, it was treated as the only mechanism of inheritance, as if the only way for offspring to resemble to their parents was by sharing genes. This is patently false, but it has taken until now for evolutionists to return to their roots by thinking in terms of heredity, with genes as one mechanism of inheritance.

This opens the door for studying the human capacity to transmit learned information across generations as both a product of genetic evolution and an evolutionary process with its own inheritance mechanisms (see my "One Culture" essay and interview with Eva Jablonka[23] for more).

To summarize, if large swaths of the biological and health sciences still need to absorb the meaning of Tinbergen's four questions, then the human related academic disciplines need to even more so. But wait! There's more!

3.4. Beyond the Ivory Tower

Beyond the Ivory Tower, there are legions of politicians, economists, and policy experts of all stripes charged with managing our affairs and making our world a better place. Most are college educated in topic areas that developed without reference to evolutionary theory since before they were born, however sophisticated these topic areas became in other respects. Some (especially politicians) are beholden to constituents who know even less, such as the roughly 50 percent of Americans who claim not to accept evolution and the other 50 percent who nominally accept it but who know scarcely anything about what this might mean for understanding and improving the human condition. If Tinbergen's four questions are needed inside the Ivory Tower, then they are needed outside the Ivory Tower even more so!

---

[23] [https://web.archive.org/web/20170502004051/https://evolution-institute.org/articl e/beyond-genetic-evolution-a-conversation-with-eva-jablonka/.]

Improving literacy might seem like a hopeless quest, but I am actually optimistic, based on extensive experience since co-founding the Evolution Institute in 2008. Here I will report progress on three fronts: Social Darwinism, Evonomics, and Prosociality.

Social Darwinism: One reason that is often cited for the rejection of evolutionary theory in relation to human affairs is "Social Darwinism," or the use of the theory to justify social inequality. Here is where a fine-grained analysis of scientific and social history, written in a way that is accessible to a broad audience, becomes essential, which TVOL has provided in a series of articles titled "Truth and Reconciliation for Social Darwinism."[24]

It turns out that the term "Social Darwinism" has always been used as a pejorative for "laissez-faire," or the justification of the status quo, much as it is used today. For example, when the U.S. Democratic President Barack Obama called Republican social policies "Social Darwinist," he was saying that the policies benefit the rich at the expense of the poor, not that Republicans were relying upon evolutionary theory to support their claims. That would be ludicrous, since many Republicans profess to be creationists.

Based on scholarship that already exists and merely needs to be made accessible to a broad audience, we can say with authority that Darwin's theory did not unleash a plague of toxic social policies justifying inequality. At most, it was added to a quiver that was already full of other arrows, including religious and economic justifications. The claim that Darwin's theory was used to justify Hitler's war policies, either directly or indirectly, is demonstrably false, as recounted in an article featuring the work of the historian of science Robert J. Richards entitled "Was Hitler a Darwinian? No! No! No!"[25]

At the same time, Darwin clearly did influence progressive social reformers such as John Dewey, as I discuss with the philosopher

---

[24] [https://web.archive.org/web/20170921203425/https://evolution-institute.org/articl e/truth-and-reconciliation-for-social-darwinism/.]

[25] See R.J. Richards, *Was Hitler a Darwinian? Disputed Questions in the History of Evolutionary Theory.* Chicago: University of Chicago Press, 2013.

Trevor Pearce in an interview titled "Was Dewey a Darwinian? Yes! Yes! Yes!"[26] But Dewey and other figures who used evolutionary theory to argue for social equality and mutual aid, including Peter Kropotkin, Thomas Huxley, and Darwin himself, are never called Social Darwinists. Go figure. The very enterprise of making an historical figure such as Darwin and his theory morally culpable for later applications is deeply problematic. Are chemists morally accountable for the use of their compounds in Hitler's gas chambers? As with other truth and reconciliation processes, this series of articles clears the air and enables a more positive and constructive exploration of social policy from an evolutionary perspective to take place.

Evonomics: The economics profession exists both inside and outside the Ivory Tower. Inside, it is supported by a mathematical edifice that looks impressive but is based on absurd assumptions about human abilities and preferences (*"Homo economicus"*) and the social environment in which economic transactions take place (e.g., that it is at equilibrium). The economist Paul Romer calls this "mathiness"[27] and observes that it is used as a front to disguise ideology as science. The main competing school of thought, behavioral economics, challenges the orthodoxy on empirical grounds but has no theoretical framework of its own, resulting in a long list of results that appear anomalous and paradoxical against the background of orthodox theory. Some academic economists have concluded that there will be no replacement for orthodox theory. The Age of Big Theories is over and all we can do is play with data using whatever perspective we like.

Outside the Ivory Tower, policies are justified by appealing to iconic figures such as Adam Smith, John Maynard Keynes, Friedrich A. Hayek, and Milton Friedman in ways that bear little if any relationship to their actual work. Adam Smith's invisible hand metaphor is invoked in ways that would make him cringe. The fact that the

---

[26] [https://web.archive.org/web/20170621234745/https://evolution-institute.org/articl e/was-dewey-a-darwinian-yes-yes-yes-an-interview-with-trevor-pearce/.]

[27] [https://paulromer.net/mathiness/.]

novelist Ayn Rand is cited as much as the economic icons adds proof, if more is needed, that the entire show is driven by purpose-driven narratives, not science.

Again, I am optimistic that something can be done, despite the enormity of the problem, based on progress that has already been made. One key is to realize that progress needs to be made both inside (the science) and outside (the narrative) the Ivory Tower, with the science and narrative connected to each other more responsively and responsibly than in the past. On the science end, the Age of Big Theories is not over. When we step away from the encapsulated world of academic economics, we can see that two Big Theories are alive and well. The first is complex systems theory, which explains the dynamics of complex systems of all sorts. The second is evolutionary theory, which explains the dynamics of living systems of all sorts. Together, they provide a new scientific foundation for economics. My colleagues and I are communicating this message to academic economists through workshops, conferences, and their published outputs. For example, the lead article for a special issue of the *Journal for Economic and Behavior Organization* entitled "Evolution as a General Theoretical Framework for Economics and Public Policy"[28] is explicitly framed in terms of Tinbergen's four questions, and an edited volume entitled *Complexity and Evolution: Toward a New Synthesis for Economics,* based on a five-day conference organized with the help of Germany's Ernst Strüngmann Forum, was recently published by MIT Press.[29]

Explaining all of this to a large and diverse audience outside the Ivory Tower, including the lay public, in addition to politicians, applied economists, and policy experts of all stripes, is easier than you might think. After all, Tinbergen's four questions are far, far

---

[28] D.S. Wilson and J.M. Gowdy, "Evolution as a General Theoretical Framework for Economics and Public Policy," *Journal of Economic Behavior and Organization*, 2013, **90**(Supplement): S3–S10. [https://www.sciencedirect.com/science/article/abs/pii/S016726 8112002673?via%3Dihub.]

[29] D.S. Wilson and A. Kirman, eds., *Complexity and Evolution: Toward a New Synthesis for Economics.* Cambridge, MA: MIT Press, 2016.

easier to understand than orthodox economic theory! In addition to content on TVOL and SEF, the EI has assisted in the creation of Evonomics.com,[30] which has achieved a circulation of a quarter million page views per month during its first year, including many thought leaders among its authors and readers. Check it out and judge for yourself how we are doing on the narrative front. My own assessment is that Tinbergen's four questions can be intuitive, appealing, and eminently useful on both sides of the Ivory Tower.

Prosociality: It is ironic that Social Darwinism is associated with the justification of ruthless competition, because the truth is just the reverse. A proper understanding of evolution in relation to human affairs reveals the importance of prosociality and identifies practical strategies for achieving it at scales ranging from small groups to the global village. Prosociality can be defined as any attitude, behavior, or institution oriented toward the welfare of others or one's group as a whole. It therefore includes behaviors that are labeled "cooperative" and "altruistic."

Here are some examples of Tinbergen's four questions in action on the topic of prosociality. Starting with the function question, as a basic matter of tradeoffs, behaving for the good of one's group requires time, energy, and risk on the part of group members, which makes prosocial individuals vulnerable to passive free-riding and active exploitation by less prosocial individuals. This should be true for all social species, although the details will depend upon the ecological context of cooperation and the properties of the species (the history question, highlighting the importance of cross-species comparisons). A large family of models has been built and tested by behavioral ecologists, along the lines that I described for optimal foraging theory. E.O. Wilson (1975) called altruism "the central problem of Sociobiology" and six chapters of the *Introduction to Behavioural Ecology* (fourth edition) are devoted to various forms of prosociality.

---

[30] [http://evonomics.com/.]

Even though prosocial behaviors can evolve in theory and many cases have been documented, the same can be said about antisocial behaviors. This is a hard lesson to learn about nature. Many animal societies are highly despotic and unjust in human terms. Individuals stay in groups because to strike out on one's own would be even worse. Sexual conflict can be severe, with males pursuing their reproductive success in ways that are highly detrimental to females and vice versa, all depending upon the circumstances. In some species, the highest source of mortality on infants is members of the same species, killing the babies of others so that they can have their own.

The selection pressures favoring prosocial and antisocial behaviors are not static but can themselves evolve. When mechanisms evolve that suppress the advantages of disruptive self-serving competition within groups (the mechanism question), prosociality evolves to such an extent that the group becomes a higher-level organism. All of the entities that we currently recognize as individuals, such as single-celled and multicellular organisms, evolved not by small mutational steps from other individuals, but as highly regulated social groups of lower-level units—a process called a "major evolutionary transition." This is something that was beyond Darwin's imagination.

The capacity to communicate with symbols and transmit large amounts of learned information (the mechanism question) is a form of prosociality and also an evolutionary process in its own right. This means that Tinbergen's four questions need to be asked for products of cultural evolution, no less than for products of genetic evolution. When human groups became bigger with the advent of agriculture, they outstripped our genetically evolved ability to enforce prosociality and became despotic, ironically more like chimpanzee societies than small-scale human societies (the history question). Despotic groups tend to fare poorly in intergroup competition with more internally prosocial groups, however, so cultural group selection, driven largely but not entirely by warfare, led to the mega-societies of today. Modern nations vary considerably in their degree of prosociality, expressed both internally and toward other nations. The same genetic and cultural evolutionary forces that brought us to our current condition are still operating.

Tinbergen's four questions have been especially insightful for the study of religion. Religions puzzle the scientific imagination because they seem both irrational and counterproductive. It's easy to understand why people would *make* blankets, but why would they *burn* them in sacrifice to imaginary agents for which there is no empirical evidence?

Two broad solutions have been proposed from the beginning of scholarship on the subject. First, religious beliefs and practices might be just as irrational and counterproductive as they seem and persist as byproducts of other beliefs and practices that are useful in non-religious contexts. Second, religious beliefs and practices might have a hidden logic and utility after all. Émile Durkheim was a proponent of the latter view, which is reflected in his famous definition of religions as:[31]

> ... a unified system of beliefs and practices relative to sacred things ... which unite into one single moral community called a Church, all those who adhere to them.

However, more than 150 years of scholarship and analysis of religion have not resolved the issue. The study of religion from a modern evolutionary perspective started at the turn of the twenty-first century and has made impressive progress reaching a consensus. Religions are a fuzzy set comprising many beliefs and practices. As with all products of evolution, adaptations come with byproducts, but the idea that religions writ large are byproducts can be authoritatively rejected. Instead, most enduring religions are impressively designed to foster prosociality among members of the religious community, which is what Durkheim proposed, although the evolutionary per-

---

[31] É. Durkheim, *The Elementary Forms of the Religious Life*. London: George Allen & Unwin, 1915; numerous editions. (Originally published as *Les formes élémentaires de la vie religieuse* in 1912.) (The passage cited occurs near the end of Book 1, "Preliminary Questions," Chapter 1, "Definition of Religious Phenomena and of Religion," on p. 62 of the 1965 Free Press paperback edition—Ed.)

spective avoids many of the problems that became associated with the tradition of functionalism that he initiated.

A common distinction between the "vertical" and "horizontal" dimensions of religion maps intriguingly onto Tinbergen's "function" and "mechanism" questions. While religions are consistently designed to foster prosociality within religious groups, interactions among groups can range from highly prosocial to highly antisocial, depending upon the ecological conditions.

James Coan's research described earlier takes the study of prosociality in a mechanistic direction. The line of research that led him to read *An Introduction to Behavioural Ecology* involved placing people in a brain scanner, stressing them with the possibility of an electrical shock, and measuring the effect of holding the hand of a loved one on the brain's response to the stress. Holding hands had a huge calming effect. In a second experiment, the loved one outside the scanner was the one being stressed with the possibility of an electric shock and the person inside the scanner was the one being supportive by holding hands. According to Coan, a skilled reader of brain scans would be required to tell the difference between someone being shocked and someone holding the hand of a loved one being shocked. Thanks to his newly acquired, four-question perspective, Coan now thinks of the brain as an organ that seamlessly integrates personal resources and social resources in making its trade-off decisions.

Research on prosociality can also be taken in a developmental direction. It is common for developmental psychologists to regard nurturing environments as optimal and harsh environments as debilitating, resulting in various pathologies comparable to a car breaking down. From an evolutionary perspective, nearly all species occupy a range of environments, from benign to harsh, during their evolutionary histories and are adapted to respond appropriately to the full range. Facultative responses to harsh environments include discounting the future and withholding prosociality toward others, since prosociality is not being extended to oneself. To say that these responses are facultative does not mean that they are voluntary. Many responses take place beneath conscious awareness and some aren't even cognitive in nature, such as an acceleration of sexual maturation. Developmental responses to harsh environments can begin before birth, can be difficult to reverse during the lifetime of the organism, and can be transmitted to offspring through epigenetic in addition to social inheritance mechanisms.

Public health efforts tend to be highly fragmented, with every problem considered in isolation. An evolutionary perspective helps us to see that prosociality is a master variable: Having it results in multiple assets and not having it results in multiple liabilities. If there could be only one policy prescription, it would be to foster prosocialilty.[32] The way to do this is to construct social environments that allow prosociality to succeed as a behavioral strategy in competition with less prosocial strategies. This can be done in a very practical way by implementing a number of core design principles that provide a strong group identity, prevent disruptive self-serving behaviors within the group, and cultivate appropriate relations with other groups. The Evolution Institute has developed a framework for coaching groups in the core design principles called PROSOCIAL. Please visit *PROSOCIAL Magazine* to learn more, including case studies of groups that have improved their efficacy by implementing the core design principles.

3.5. Advice for a College Applicant

Earlier I stated that the future is already here for a growing number of scientists and scholars who employ the four-question approach in all topic areas. In the spirit of AcademicInfluence.com's primary mission, here is some advice for students who wish to receive a fully rounded, four-question college education.

---

[32] A. Biglan, *The Nurture Effect: How the Science of Human Behavior Can Improve Our Lives and Our World*. Oakland, CA: New Harbinger Publications, 2015.

Now is a good time to start. Neo-Darwinism is not an advanced topic. Any inquiring person can learn the basics, especially when the relevance of neo-Darwinism to matters of importance is made clear. As you have probably gathered from reading this essay, once learned, Tinbergen's four questions can help you understand nearly everything that interests you in your personal and future professional life.

Don't expect it to come automatically. Academic culture can be surprisingly conservative. A decade or more can be required for new developments in science and scholarship to find their way into textbooks. At the vast majority of colleges and universities worldwide, evolution is still taught as a biological topic. Even if you become a biology major, many of your courses will focus on mechanisms in a way that crowds out the other three questions. If you major in a human-related topic, most of your professors will have had little or no training in evolution during their own higher education, although some will have picked it up on their own.

The best education is self-education. Given this somewhat bleak assessment, it's up to you to seek out a four-question education. The more you can learn on your own, the better you will be able to direct your search. I wrote my book *Evolution for Everyone: How Darwin's Theory Can Change the Way We Think About Our Lives* (Delacorte Press, 2007)[33] with this in mind. Chapter 10, which begins on page 63, is titled "Your Apprentice License," which means that I was able to provide the basics in the previous nine short chapters. The remaining 26 chapters apply the basics to a smorgasbord of biological and human-related topics. From there (or even before), I suggest perusing the articles on This View of Life, which cover "anything and everything from an evolutionary perspective." Then just follow your bliss. Whatever interests you, it is likely that others have started

---

[33] [https://www.amazon.com/Evolution-Everyone-Darwins-Theory-Change/dp/0385340923/ref=sr_1_1?crid=2SX8EKONZBTGQ&keywords=david+sloan+wilson+evolution+for+everyone&qid=1662996606&s=books&sprefix=david+sloan+wilson+evolution+for+everyone%2Cstripbooks%2C80&sr=1-1.]

to approach it from an evolutionary perspective—although there will still be plenty of uncharted territory for you to explore!

Seek out the professors who are at the forefront of what interests you. Once you have refined your own interests, you can check out the colleges and universities that include the leaders in those areas. Are you aiming for medical school? Then check out Arizona State University, home of the Center for Evolution and Medicine,[34] or a consortium of four universities in North Carolina that have created the Triangle Center for Evolutionary Medicine.[35]

Are you fascinated by cultural anthropology? Try UCLA, UC Davis, or Harvard. The Arts? Check out the NeuroArts Lab[36] at McMaster University in Ontario, Canada. Don't be shy about emailing professors directly. Most of them are dedicated teachers who will be impressed by your initiative and willing to take a few moments to advise you. Tell them that you have become interested in evolutionary theory and in their research area. Ask for their opinion about their institution as a place where an undergraduate student can get a good evolutionary education and what other institutions they might recommend. Ask about opportunities for working with them and/or their graduate students.

Check out EvoS programs and their equivalents. Only a handful of colleges and universities feature a campus-wide evolutionary studies program for undergraduate students and most of them are based on the EvoS program that I started at Binghamton University in 2003. If you came to Binghamton, you could earn a certificate in Evolutionary Studies in parallel with any major. You could take our 100-level "Evolution for Everyone" course during your first semester to learn the basics. Then you could choose from a menu of other courses to deepen your knowledge in the topic areas that interest you.

You could join the EvoS undergraduate student club to socialize with like-minded individuals and with graduate students in the EvoS

---

[34] [https://evmed.asu.edu/.]

[35] [https://tricem.org/.]

[36] [https://neuroarts.org/.]

Graduate Student Association. You would find it comparatively easy to start working with EvoS faculty participants and their graduate students. You would be required to take the two-credit "Current Topics in Evolutionary Studies" seminar at least twice. This course is built around the campus-wide EvoS seminar series, which brings approximately 10 speakers to campus every semester to speak on "anything and everything" from an evolutionary perspective.

In preparation for each seminar, EvoS students are required to read one or more articles from the primary literature, write a commentary due before the seminar, attend the seminar, and attend an extended discussion with the speaker following the seminar. Students who earn the EvoS certificate will have repeated this experience 20 different times for different topics, along with the four-credit courses that they have taken and their research experiences—all in parallel with their conventional major.

Please visit the EvoS Consortium website for a list of colleges and universities that offer multicourse programs such as the one that I direct at Binghamton, along with programs under development. I think you can see how such programs create synergies that go beyond individual professors and their research programs. But don't take my word for it—here is how one EvoS student at Binghamton named Benjamin Seitz expressed the value of his education.[37]

> Ultimately, what the EvoS program does here is provide undergraduates with the opportunity to function as graduate students. Every week, we are exposed to researchers from around the world, who share a similar passion for evolutionary studies. Not only do these speakers broaden our exposure to current research in the field, but they provide phenomenal networking opportunities for those of us wishing to pursue careers in academia. On top of this, the EvoS program encourages self-learning and the pursuit of knowledge for the sake of pursuing knowledge. The evolutionary toolkit, which is so feverishly promoted by the EvoS program, is a phenomenal catalyst

---

[37] [https://evolution-institute.org/blog/a-student-expresses-the-value-of-an-evolutionary-education/.]

for stimulating intellectual discussion. It is a tool and a perspective that once turned on, is seemingly impossible to turn off.

Lastly, the EvoS program provides a community on campus. When I was in high school and looking at colleges, I was torn between the intimacy of small liberal arts colleges and the vast possibilities offered at large research intuitions. I knew I wanted to get involved in research, but I figured the best way to do so would be to go to a tiny college where I could get to build strong relationships with my professors, and perhaps have a beer with them if I was lucky. Fortunately, I ended up at a large research institution, and the intimacy that I was seeking from a small college I found immediately from the EvoS program. It's a rather simple formula really: bright and welcoming people, who share a common interest and understanding of the world, who are highly encouraging and respectful to all those interested in getting involved. That's how I see the EvoS Program here at Binghamton, and this program will by far be what I cherish most about my four years here.

3.6. Is Neo-Darwinism Enough for the Experts?

So far, I have made the case that the Neo-Darwinian revolution is far from complete and that its completion will be transformative for both academic knowledge and our practical ability to make the world a better place. Against this background, it is a distraction to say that neo-Darwinism is not enough.

Some readers might be willing to grant me this claim but might still argue that neo-Darwinism is not enough for the experts at the cutting edge of evolutionary science. In this final section of my essay, I will argue that neo-Darwinism is enough for the experts. My argument will be consistent with enthusiastic support for the so-called "extended evolutionary synthesis" (EES).[38]

---

[38] K.N. Laland, T. Uller, M.W. Feldman, K. Sterelny, G.B. Müller, A. Moczek, E. Jablonka, and J. Odling-Smee, "The Extended Evolutionary Synthesis: Its Structure,

Terms such as "paradigm," "research program," and "synthesis" are used loosely by scientists and the lay public, often in ways that are self-promoting, but they are also used in more careful ways by historians and philosophers of science to understand continuities and discontinuities in the scientific process. Massimo Pigliucci, who has a PhD in both philosophy of science and evolutionary biology, carefully coined the term "extended evolutionarysSynthesis" to signal continuity. He described it to me this way during a 2016 interview.[39]

> Now for the key concept of "synthesis." This is not a philosophical term [in contrast to "paradigm" and "research program", as it was introduced by [Julian] Huxley with the title of his famous book (Evolution: The Modern Synthesis) [1942]. What he meant to convey was the idea that the MS was not something radically different from Darwinism and neo-Darwinism (the late 19th century modification of the original theory that got rid of Lamarckian influences, largely thanks to the work of August Weissman and Alfred Wallace). Rather, it was a merging, a reconciliation, of Darwinism with the new discoveries coming out of genetics, and in particular the demonstration, achieved by Ronald Fisher, that Darwinism and Mendelism were not at all at odds with each other, as many thought at the time. The Synthesis then got expanded to a number of additional disciplines, from natural history to zoology and botany, and of course to paleontology (but, crucially, not to embryology and developmental biology).
>
> It is in this same sense that most proponents of an Expanded Synthesis use the term: we don't think that we are witnessing a Kuhnian paradigm shift, or the replacement of a Lakatosian research program by another one. We are, however, in need of explicitly and organically incorporating into the framework of the MS a number of new discoveries and concepts (phenotypic plasticity, epigenetic inheritance,

---

Assumptions and Predictions," *Proceedings of the Royal Society of London B*, 2015, **282**(1813): 20151019. [http://dx.doi.org/10.1098/rspb.2015.1019.]

[39] [https://web.archive.org/web/20170503014127/https://evolution-institute.org/articl e/the-origin-of-the-extended-evolutionary-synthesis-an-interview-with-massimo-pigliucci/ .]

evolvability, and so forth) that were unknown to, or unappreciated by, the architects of the Modern Synthesis.

Like Richard Lewontin, Pigliucci appreciates the personal and sociological aspects of science in addition to the intellectual aspects.

> We tend to forget that science is a human enterprise, and as such— at the least in the short run—affected by social dynamics and power struggles. One of the dominant personalities during the period in which the MS congealed was Ernst Mayr, who staunchly defended a number of notions that were important to him (such as allopatric speciation), and equally forcefully rejected others that didn't fit his view of evolution (such as G.G. Simpson's distinction between bradytelic and tachytelic evolution). It was Mayr who famously justified the exclusion of developmental biology from the Synthesis on the basis that, allegedly, developmental biologists were simply not interested in evolution. In fact, many were, but subscribed to views more similar to those of the famous geneticist Richard B. Goldschmidt, who proposed the idea of "hopeful monsters" to account for speciation and for major transitions in the fossil record. That idea didn't sit well with Mayr's emphasis on gradualism, and was accordingly purged from the canon, despite Goldschmidt's stellar reputation at the time. Today we think that the notion of hopeful monsters was indeed misguided, but also that Goldschmidt was more prescient than Mayr in understanding the fruitful interaction between genetics and developmental biology—something that nowadays goes under the name of "evo-devo."

The evolutionary biologist Kevin Laland has taken a leadership role in developing the EES, including a major grant from the John Templeton Foundation that will fund over 20 research projects. He also emphasizes continuity in my 2016 interview with him,[40] from which the following is taken.

> I was drawn to thinking about these issues through my research on niche construction, with John Odling-Smee (Oxford) and Marc

---

[40] [https://web.archive.org/web/20170707144816/https://evolution-institute.org/articl e/empowering-the-extended-evolutionary-synthesis/.]

Feldman (Stanford). John was a participant at the Altenberg meeting (organized by Pigliucci), and he and I, together with evolutionary ecologist Tobias Uller (then Oxford, now Lund), began discussing whether there was a conception of the EES that could do useful work. We had no sympathy with the argument that evolutionary biology was undergoing a "paradigm shift"—to my mind paradigm shifts are an outdated notion (sciences change more through gradual evolution than dramatic revolution). Nonetheless, we were very conscious of how academic fields possess conceptual frameworks —ways of thinking—that influence what questions are asked, what data is collected, and how that data is interpreted. Here, alternative perspectives can be of real value to the extent that they encourage researchers to generate and test novel hypotheses, or open up new lines of inquiry. That is how we envisaged the EES could be of service —as an alternative way of thinking about evolution, which could be deployed alongside traditional perspectives to stimulate innovative research.

However, any such alternative needs to be formulated in a disciplined way. We noted that certain literatures—for instance, those concerning developmental bias, developmental plasticity, and expanded views of inheritance—stood out as being open to both traditional and progressive interpretations, leaving them lying squarely on the fault line. In all cases, the more progressive reading emphasized an organism-centered perspective, rejected the idea that development was controlled by a genetic program, and recognized that developmental processes played important (and not fully appreciated) evolutionary roles. It made sense to conceptualize the EES as an eco-developmental perspective, and to highlight these literatures as the intellectual territory on which an EES might focus.

None of these considerations require going beyond Tinbergen's four questions, although they do highlight the relative neglect of the "development" question during the middle of the twentieth century, which began to be remedied with the advent of the "evo-devo" movement (this term was coined in the 1990s).

Here are three examples of major developments in evolutionary science that still fall squarely within neo-Darwinism as I have defined it. I will focus on my own research projects, not because I think I made outsized contributions, but because of my familiarity with them. If my research can be called distinctive, it is by showing how someone equipped with Tinbergen's four questions can parachute into nearly any topic area and make a contribution as measured by the gold standard of expertise in science—peer-reviewed publications.

Multi-level selection. George C. Williams started to write *Adaptation and Natural Selection* (1966) in the late 1950s to correct what he regarded as sloppy thinking about evolution. His PhD training at UC Berkeley included population genetics, which made it second nature for him to think about natural selection in terms of relative fitness. It doesn't matter how well an organism survives and reproduces in any absolute sense, only that it does so better than other organisms in its vicinity. Relative fitness creates the problem that group selection is needed to solve. As a basic matter of trade-offs, the traits that maximize relative fitness within a social group are unlikely to benefit the group as a whole. Traits that are "for the good of the group" require time, energy, and expense on the part of group members that can be exploited by passive free-riding and active exploitation by other group members. For these traits to evolve, there must be a process of selection among groups in a multi-group population, which is largely opposed by selection among individuals within groups.

Darwin understood this, but many scientists during the first half of the twentieth century, including ecologists who were newly learning about evolution, naively assumed that adaptations could evolve at the level of individuals, groups, species, and ecosystems without needing to distinguish among the levels. When the need for multiple levels of selection was acknowledged, it was often assumed that higher-level selection easily trumped lower-level selection. As one of the most widely used ecology textbooks of the 1950s put it, the prevailing law of nature was "all for one and one for all."[41]

In his effort to combat this kind of naïve adaptationism, Williams asserted that it was most parsimonious to invoke individual-level selection and to require proof for putative examples of higher-level selection. Based on his own review, he concluded that higher-level selection was almost invariably trumped by lower-level selection, so that "group-level adaptations do not, in fact, exist." If individuals appear to behave for the good of their groups, then their apparent altruism needs to be explained as a form of self-interest.

By the time *Adaptation and Natural Selection* was published in 1966, W.D. Hamilton had published his theory of inclusive fitness,[42] which showed how individuals could evolve to benefit "their" genes in the bodies of their genetic relatives. Then, Robert Trivers introduced the concept of "reciprocal altruism" in 1971,[43] which explained apparent altruism in terms of return benefits to the altruist. These two new theoretical frameworks, coupled with Williams's skeptical analysis of group selection and invocation of parsimony, made group selection appear as dead within evolutionary theory as Lamarckism. When I began graduate school that year, it was almost mandatory for authors to assure their readers that they were not invoking group selection.

But that is also when the tide began to turn. An obscure theoretical biologist named George R. Price created an equation that statistically partitioned natural selection in a multi-group population into within- and between-group components. When W.D. Hamilton compared his theory of inclusive fitness to the Price Equation, he was shocked to discover that they were equivalent.[44] His own formulation obscured

---

[41] W.C. Allee, A.E. Emerson, O. Park, T. Park, and K.P. Schmidt, *Principles of Animal Ecology.* Philadelphia: Saunders, 1949.

[42] W.D. Hamilton, "The Genetical Evolution of Social Behavior: I and II," *Journal of Theoretical Biology,* 1964, **7**: 1–52.

[43] R.L. Trivers, "The Evolution of Reciprocal Altruism," *Quarterly Review of Biology,* 1971, **46**: 35–57.

[44] For a book-length account, see O.S. Harmon, *The Price of Altruism.* New York: Norton, 2010.

the fact that when social interactions take place among genetic relatives, there are multiple groups. Sometimes, the groups are spatially defined, as when a butterfly lays a clutch of eggs on a leaf. Sometimes, the groups are behaviorally defined, as when an individual preferentially interacts with a relative and avoids interacting within non-relatives. Either way, the evolving population is composed of multiple local groups as far as the social interactions are concerned. This fact, which might seem obvious in retrospect, was made clear to Hamilton by the Price Equation, which also showed that altruism is selectively disadvantageous within kin groups and requires selection among kin groups to evolve. Hamilton wasn't wrong about genetic relatedness as an important factor for the evolution of altruism, but he was wrong to regard his theory as an alternative to group selection, as he announced to the world in a 1975 article.[45]

That was also the year that I published my first article on group selection,[46] which noted that social interactions almost invariably take place among sets of individuals that are small compared to the total evolving population. If these sets of individuals (I called them "trait-groups") are regarded as the groups of a multilevel selection model, then traits conceptualized as only "apparently" altruistic in kin selection and reciprocal altruism models are "really" altruistic, in the sense that they are selectively disadvantageous within trait-groups and require selection among trait-groups to evolve. My algebraic model was different from Price's statistical model, but both made the same basic point.

A very important new concept for the history and philosophy of science was emerging that has become known as "equivalence."[47] In the standard portrayal of science, alternative hypotheses invoke

[45] W.D. Hamilton, "Innate Social Aptitudes in Man: An Approach from Evolutionary Genetics," in R. Fox, ed., *Biosocial Anthropology*. London: Malaby Press, 1975; pp. 133–155.

[46] D.S. Wilson, "A General Theory of Group Selection," *Proceedings of the National Academy of Sciences, USA*, 1975, **72**: 143–146.

[47] For an accessible review, see Chapter 3 of D.S. Wilson, *Does Altruism Exist? Culture, Genes, and the Welfare of Others*. New Haven, CT: Yale University Press, 2015.

different causal explanations of a given phenomenon, such that one can be shown to be right and the other wrong on the basis of empirical evidence. Even paradigms are replaced by other paradigms, although the process is messier and more protracted. No one (other than historians) talks about pre-Copernican views of the solar system anymore.

But some alternative explanations are not like that. Rather than invoking different causal processes, they invoke the same causal processes from different perspectives. As such, they deserve to coexist to the extent that the different perspectives offer novel insights. Familiar analogies from everyday life include different financial accounting systems, different languages, and viewing complex objects such as a mountain from different directions.

The history of the group selection controversy from 1975 to the present is the history of confusing theories that offer different perspectives on the same causal processes with theories that invoke different causal processes. Here is the sober assessment of two philosophers of science, Jonathan Birch and Samir Okasha in a 2014 article:[48]

> In earlier debates, biologists tended to regard kin and multilevel selection as rival empirical hypotheses, but many contemporary biologists regard them as ultimately equivalent, on the grounds that gene frequency change can be correctly computed using either approach. Although dissenters from this equivalence claim can be found, the majority of social evolutionists appear to endorse it.

Why four decades were required to reach this consensus will be a juicy topic of conversation among historians of science for years to come. Personal and sociological factors were almost certainly involved, as Lewontin and Pigliucci have stressed for other aspects of evolutionary thought. It probably wasn't a coincidence that the "everything is selfish" perspective among evolutionists coincided with a similar perspective in economics and methodological individ-

---

[48] J. Birch and S. Okasha, "Kin Selection and Its Critics," *BioScience*, 2014, **65**(1): 22–32. [http://doi.org/10.1093/biosci/biu196.]

ualism in the social sciences during the second half of the twentieth century. In any case, the entire controversy is primarily a clarification of Tinbergen's "function" question and does not require going beyond the four-question approach.

Complex systems theory. In his book *Chaos: Making a New Science* (Viking, 1987),[49] James Gleick described how the study of complex systems was retarded by theorists who regarded analytical mathematical models as superior to other modeling approaches, such as computer simulations—and therefore ignored interactions that were too complex to model with analytical equations. For this and other reasons, the formal study of complex systems is remarkably recent. The institute most closely associated with complex systems theory, the Santa Fe Institute, was founded in 1984.

Complex systems theory can be said to be more general than evolutionary theory because it covers complex systems of all sorts, living and non-living. Pioneers of complex systems theory include some biologists, such as Robert May, whose work on chaotic population dynamics is described in Gleick's book, but also many non-biologists specializing in the study of complex physical systems such as the weather or computer simulation models such as English mathematician John H. Conway's "Game of Life,"[50] in which "agents" following simple rules interact to produce an amazing variety of system-level behaviors.

Complex systems theory has profound consequences for all four of Tinbergen's questions. On the other hand, complex systems theorists who are not biologists are likely to be naïve about Tinbergen's four questions, however brilliant they may be in other respects. I mean no disrespect by making this statement. I am merely observing that the same process of integration that took place in topic areas such as animal behavior, ecology, and population genetics decades

---

[49] [https://www.amazon.com/Chaos-Making-Science-James-Gleick/dp/0143113453/ref=sr_1_1?crid=3TWSTAQ499I29&keywords=gleick+chaos&qid=1663001371&s=books&sprefix=gleick+chaos%2Cstripbooks%2C89&sr=1-1.]

[50] [https://en.wikipedia.org/wiki/Conway%27s_Game_of_Life.]

earlier is now needed for complex systems theory. Why should it be otherwise?

The key term "complex adaptive system" (CAS) provides an example.[51] It has at least two meanings: A complex system that is adaptive as a system (CAS1); and a complex system of agents that follow adaptive strategies (CAS2). Examples of CAS1 include social insect colonies, brains, and the immune system. Examples of CAS2 include multispecies ecosystems, crowds, and the stock market. If you visit the Wikipedia entry for "complex adaptive system"[52] (or some other, more authoritative source, if you like), you will see both CAS1 and CAS2 systems lumped under a single label as if there is no need to distinguish between them. This signals confusion about units of functional organization—the very problem that is addressed by multilevel selection theory.

Complex interactions can produce lots of *pattern* at the system level, but they are no more likely than a point mutation to produce *adaptive* pattern without a process of selection. Once this foundational point is grasped, then we can proceed to study how multilevel selection operating on complex systems differs from multilevel selection operating on more simple systems. For example, in models that assume a simple relationship between genes and phenotypic traits (e.g., a single genetic polymorphism that codes for altruism vs. selfishness), phenotypic variation among groups is directly proportional to genetic variation among groups. If groups are formed by drawing N individuals at random from a large population, then the amount of genetic and phenotypic variation among groups will decline rapidly with the value of N as an inevitable consequence of sampling error. This is one reason why genetic relatedness (= groups initiated by a small number of individuals) is considered so important for the evolution of altruism.

---

[51] D.S. Wilson, "Two meanings of complex adaptive systems," in D.S. Wilson and A. Kirman, eds., *Complexity and Evolution: A New Synthesis for Economics*. Cambridge, MA: MIT Press, 2016.

[52] [https://en.wikipedia.org/wiki/Complex_adaptive_system.]

Now, let's assume a more complex genotype-phenotype relationship. Instead of coding directly for a given phenotypic trait, genes code for component traits that interact with each other to produce the phenotypic trait. With this alteration of the model, something magical happens. *Genetic* variation among groups declines with N, as before, but *phenotypic* variation among groups can remain high, even with very large values of N. This is because even tiny initial differences among groups don't remain tiny, but can be magnified by complex interactions taking place within each group. This is called "sensitive dependence on initial conditions" and also explains why the weather is unpredictable and how a small genetic change in an organism (such as a single-nucleotide substitution) can be magnified by developmental processes to result in a large phenotypic change.

The bottom line is that selection at the level of higher-level units such as single-species groups and multi-species communities might be much more potent than expected on the basis of simple models. To test this hypothesis, William Swenson and I created microbial ecosystems in the laboratory that were initiated by many hundreds of species and millions of individuals from a single well-mixed source.[53] Initial variation among the ecosystems, based on sampling error, was vanishingly small. Nevertheless, these differences did not remain small, but grew larger during the course of a few days and weeks (= many microbial generations) based on complex interactions taking place within each ecosystem. The differences in species composition and the genetic composition of each species influenced measurable properties of the ecosystems, such as pH or the ability to degrade a toxic compound. When ecosystems were selected on the basis of these properties and used colonize a new set of ecosystems, there was a response to selection, which is proof of ecosystem-level heritability. In short, we demonstrated that ecosystems can be selected for their

---

[53] W. Swenson, J. Arendt, and D.S. Wilson, "Artificial Selection of Microbial Ecosystems for 3-chloroaniline Biodegradation," *Environmental Microbiology*, 2000, 2: 564–571; and W. Swenson, D.S. Wilson, and R. Elias, "Artificial Ecosystem Selection," *Proceedings of the National Academy of Sciences, USA*, 2000, **97**: 9110–9114.

properties in much the same way as individuals, which appeared theoretically impossible based on simpler models.

This research was published in 2000 as two articles in journals that are hard to ignore (*Proceedings of the National Academy of Sciences* and *Environmental Microbiology*)—but nevertheless they were largely ignored. Many evolutionists were still in denial about multilevel selection and many microbiologists focused on mechanisms to the exclusion of Tinbergen's other three questions. The fact that our inquiry was led by Tinbergen's "function" question and could inform the study of mechanisms did not impress such microbiologists—including the reviewers of grants that we submitted to continue the research.

Two thousand was also the year that the term "microbiome"[54] started to be used with increasing frequency. By now, it is widely appreciated that every multicellular organism is inhabited by an ecosystem of microbes and other small creatures in numbers that often exceed that of the organism's own cells. Selection at the level of the microbiome has become impossible to ignore and other laboratories are beginning to continue the line of research initiated by Swenson and myself, with a greater capacity to address all four of Tinbergen's questions than we had.[55] For the purpose of this essay, the main point to stress is that complex systems theory is profoundly relevant to Tinbergen's four questions without necessitating going beyond them.

Evolution as a directed process. One of the dogmas that became established early during the history of evolutionary thought, thanks largely to August Weismann, is that phenotypic variation, while not necessarily random in the strict sense of the word, is nevertheless arbitrary and undirected with respect to the traits that are being selected. Anything that smelled of directed evolution was branded

---

[54] [https://en.wikipedia.org/wiki/Microbiota.]

[55] For example, K. Panke-Buisse, A.C. Poole, J.K. Goodrich, R.E. Ley, and J. Kao-Kniffin, "Selection on Soil Microbiomes Reveals Reproducible Impacts on Plant Function," *The ISME Journal*, 2015, **9**: 980–989; doi: 10.1038/ismej.2014.19.

with the label "Lamarckian" and declared impossible, much as group selection was declared impossible during the 1960s and 70s.

Backing away from this dogmatic position is an important part of the extended evolutionary synthesis. For me, the best way to think about directed evolution is to focus on animal behavior, which brings the arc of this essay back to Tinbergen. In the conventional view of natural selection, mutations are not directed, but they result in the evolution of behaviors that are indubitably directed. Optimal foraging behavior is anything but random or arbitrary with respect to searching for prey!

Since the work of James Mark Baldwin,[56] we have known that directed behaviors that are a product of undirected evolution can double back to influence the evolutionary process. What the organism chooses to do by learning alters the selection pressures operating on the genes of the organism. This was celebrated as a major insight at the time—a form of directed evolution that was fully consistent with Weismann's doctrine—but very little was done with it over the ensuing decades.

In the meantime, the study of behavior became increasingly sophisticated, including the distinction between closed and open forms of phenotypic plasticity. In closed forms, the behavioral alternatives are genetically programmed and triggered by environmental cues, such as a tadpole that is prepared at birth to either forage or seek a refuge, depending upon the chemical scent of its predator.

In open forms, organisms vary their behavior in a relatively open-ended fashion and adopt the behaviors that are most rewarding, such as a rat that learns to press a lever to receive food in a Skinner box. The capacity to learn in an open-ended fashion evolved by genetic evolution—for example, reinforcers such as pleasure and pain are genetically programmed, but they result in behaviors that are not

---

[56] J.M. Baldwin, "Development and Evolution," *Philosophical Review*, 1903, **12**(4): 442–451.

genetically programmed, such as the pigeons that Skinner reinforced to play ping pong.

Open forms of phenotypic plasticity are properly regarded as evolutionary processes built by other evolutionary processes, or "Darwin machines" to use the felicitous phrase coined by William Calvin and elaborated upon by Henry Plotkin.[57] Unlike genetic evolution, Darwin machines are expected to be directed forms of evolution, although they must also have an arbitrary component to remain open-ended. When Darwin machines become transgenerational through forms of social learning found in many species and forms of symbolic thought that are distinctively human, there should be no stigma whatsoever about the fact that they are partially directed.

A point I made about *An Introduction to Behavioural Ecology* as a maturation of what Tinbergen, Lorenz, and von Frisch started, is that the distinction between behavioral traits and non-behavioral traits became blurred. To most people, the length of the small intestine is not a behavioral trait. Yet, it can grow longer or shorter depending upon the foraging ecology of a species and is even a phenotypically plastic trait in some species, which means that it can become shorter or longer within an individual, depending upon what it eats. In many frog species, tadpoles don't just change their foraging behaviors depending upon the scent of a predator; they undergo a whole-body makeover.

In the same way, ideas about directed evolution that have been worked out for the study of behavior are applicable to traits that are not customarily regarded as behavioral, such as genetic and epigenetic inheritance mechanisms, the immune system, developmental programs, and neural processes in the brain. Rigid adherence to Weismann's doctrine should be declared thoroughly obsolete, along with rigid rejection of group selection. These are profound advances in evolutionary thought, but they do not require going beyond

---

[57] W.H. Calvin, "The Brain as a Darwin Machine," *Nature*, 1987, **330**: 33–34; H. Plotkin, *Darwin Machines and the Nature of Knowledge*. Cambridge, MA: Harvard University Press, 1994.

Tinbergen's four questions. If anything, they require a proliferation of Tinbergen's four questions for every evolutionary process that is built by another evolutionary process.

3.7. Conclusion

In this essay I have strived to show that the neo-Darwinian revolution is far from complete—in the biological sciences, in the human-related sciences and humanities, in economics, government, and public policy, and in higher education. Even the most advanced topics associated with the term "extended evolutionary synthesis" operate within the framework of four questions that must be addressed for all products of evolution, concerning their function, mechanism, development, and phylogeny.

Completing the neo-Darwinian revolution will be transformative and quite frankly essential for solving the problems of our age. Unless we can become wise managers of evolutionary processes, then evolution will take us where we don't want to go.[58]

Luckily, Tinbergen's four questions are sufficiently simple to learn and employ, once their relevance is understood, that widespread literacy is within reach.

---

[58] D.S. Wilson, S.C. Hayes, A. Biglan, and D. Embry, "Evolving the Future: Toward a Science of Intentional Change," *Behavioral and Brain Sciences*, 2014, **37**: 395–460.

4. Denis Noble's Major Statement: Gene-Centric Neo-Darwinism Has Failed

At the end of the Interview, I proposed that neo-Darwinism is not enough for 10 reasons. I will now explain why I listed each of those reasons.

4.1. Major Diseases Still Plague Humanity

The gene-centric view has failed in one of its major claims, i.e., that it would result, through sequencing genes, in curing the major diseases that plague humanity.

The genome is often described as the "Book of Life" by biologists favoring gene-centric views. This was one of the colorful metaphors used when projecting the idea of sequencing the complete human genome towards the end of the twentieth century. It was of course a brilliant public relations move. Who could not be intrigued by reading the Book of Life and unraveling its secrets? And who could resist the promise that, within about a decade of establishing the full draft sequence of the human genome in 2000—i.e., by around 2010—reading this "book" would reveal how to treat cancer, heart disease, nervous diseases, diabetes, and many others through the discovery of many new pharmaceutical targets?

As we all know, it simply didn't happen.[1] An editorial in *Nature* in 2010[2] (which inspired a similar editorial in *Prospect*[3]) spelled this out:

> But for all the intellectual ferment of the past decade, has human health truly benefited from the sequencing of the human genome? A startlingly honest response can be found on pages 674 and 676, where the leaders of the public and private efforts, Francis Collins and Craig Venter, both say "not much."

> The targets were identified all right. At least 200 new possible pharmaceutical targets are now known and there may be more to come, but we simply do not understand how to use them. The problem does not therefore lie in absence of knowledge about the sequences. The problem is that we neglected to do the relevant physiology. The chase to sequence everything as quickly as possible at any cost distorted the balance of health care research so much that major areas of integrative physiology are now in a very fragile state. The transmission of knowledge and skills to the next generations of researchers has become the big problem. And it requires urgent attention if we are to rescue those skills.

> The "Book of Life" represents the high-water mark of the enthusiasm with which the medical application of gene-centric Neo-Darwinism was developed. Its failure to deliver the promised advances in healthcare speaks volumes.[4]

I feel sad about this for two reasons.

First, before the shift towards genomic approaches to pharmacology, we did in fact have reasonably adequate methods for developing

---

[1] D. Noble, "The Emperor's New Genes," 2016, Institute of Art and Ideas (AIA) website. [https://iai.tv/articles/the-emperors-new-genes-auid-673.]

[2] See Editorial, "The Human Genome at Ten," *Nature,* 2010, **464**: 649–650, p. 649; see, also, M.J. Joyner and F.G. Prendergast, "Chasing Mendel: Five Questions for Personalized Medicine," *Journal of Physiology*, 2014, **592**: 2381–2388.

[3] Philip Ball, "Too Much Information," 2010, Prospect Magazine website: https://www .prospectmagazine.co.uk/magazine/too-much-information

[4] See, for example, D.G. Clayton, "Prediction and Interaction in Complex Disease Genetics: Experience in Type 1 Diabetes," *PLoS Genetics*, 2009, **5**(7): e1000540; doi: 10.1371/journal.pgen.1000540.

new drugs against specific diseases. The method was to work initially at a phenotype level to identify possible active compounds, and then to drill down towards individual protein or other molecular targets. This was the approach used so successfully by Sir James Black, the Nobel-laureate discoverer of beta-blockers and H2 receptor blockers.[5] It is the method by which the work of collaborators in my laboratory eventually led to the successful heart drug, Ivabradine.[6] But the consequence of diverting large-scale funding towards the search for new drugs via genomics has been that the Black approach is now much less common and that the pharmaceutical industry is producing fewer new medications at vastly greater cost.

Of course, the Black approach could and should be complemented by genomics, and there are successful cases where protein targets found by classical methods were later also identified as coded by particular genes. A good example is Duchenne muscular dystrophy, where the gene for the protein utrophin that can substitute, in mice at least, to cure the disease was discovered before the DNA sequence was identified.[7]

Second, although the results for healthcare are disappointing, there were very good scientific reasons for sequencing whole genomes of various species. The benefits to evolutionary and comparative biology in particular have been immense. I wish that had been the main justification for the Project. There would then have been little risk of a backlash as people come to understand the limitations in relation to healthcare. The sequencing of genomes may well eventually contribute to healthcare when the sequences can be better understood in the context of other essential aspects of physiological

---

[5] "James Black (pharmacologist)," Wikipedia article: https://en.wikipedia.org/wiki/James_Black_(pharmacologist)

[6] D. DiFrancesco and J.A. Camm, "Heart Rate Lowering by Specific and Selective I(f) Current Inhibition with Ivabradine: A New Therapeutic Perspective in Cardiovascular Disease," *Drugs*, 2004, **64**: 1757–1765.

[7] R.J. Fairclough, M.J. Wood, and K.E. Davies, "Therapy for Duchenne Muscular Dystrophy: Renewed Optimism from Genetic Approaches," *Nature Reviews Genetics*, 2013, **14**: 373–378.

function. But the promise of a peep into the Book of Life leading rapidly to a cure for all diseases was an expensive mistake. Without equally strong effort at the phenotype level, we will not reap the rewards of genome sequencing.

The evidence for the poor predictive health impact of genome sequencing compared to standard medical and physiological tests is compelling, as is shown by an online lecture by Professor Michael Joyner from the Mayo Clinic.[8]

The whole lecture is well worth watching. I will just illustrate two of the slides. The first compares the variance in human height that correlates with genomic profile, which is around 4–6%, while the variance that correlates with mid-parental height is as much as 40%.

The second illustrates the age-dependent variation of blood pressure in two populations living in island and city environments

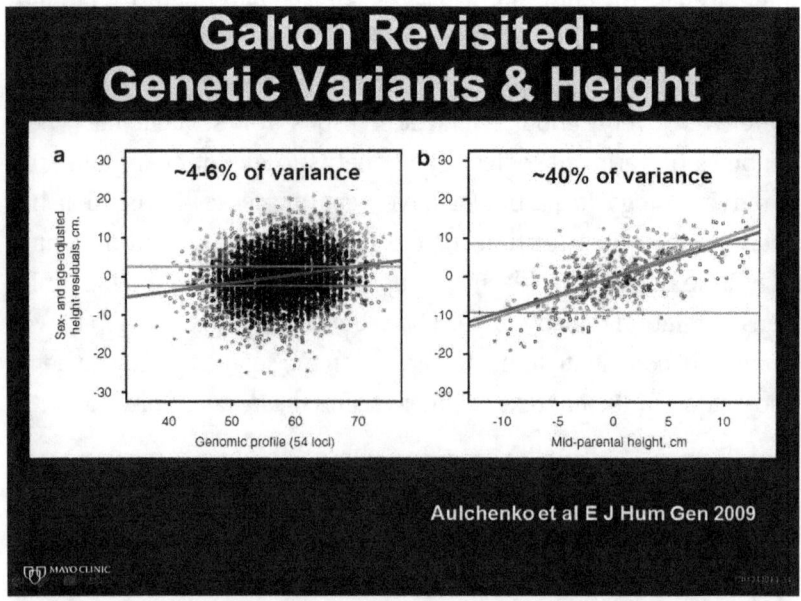

Figure 4.1: Phenotypic versus Genetic Variance – Height.

---

[8] [https://www.youtube.com/watch?v=QsOeRTa4fYs.]

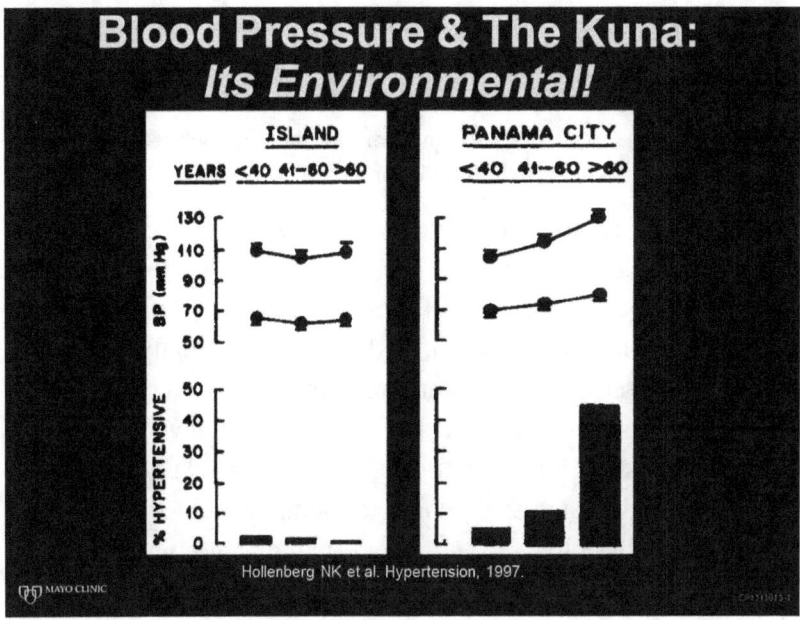

Figure 4.2: *Phenotypic versus Genetic Variance – Blood Pressure.*

in Panama. City-dwellers develop age-related hypertension. Island-dwellers do not.

These kinds of comparisons apply also to many diseases in humans. A few phenotype measurements give much more reliable prediction of disease states than do GWAS correlations. More details can be found in the videoed lecture.

4.2. Privileging Any One Level in Biological Systems Cannot Be Justified

The gene-centric view does not have a sound metaphysical basis. There is no justification for privileging any one level in biological systems. No one has ever produced such a justification.

James Watson had a go at a justification when he famously quipped "There are only molecules—everything else is sociology," which I have already referred to in the Interview. To quote my own response in the Interview:

The central idea ("there are only molecules") is what needs challenging. That is clearly not true. Why not say "there are only strings," or whatever physicists now identify as the most fundamental (note the force of this word) entities? But more importantly, why should we think that the universe cares about any particular level?

The Theory of Biological Relativity answers that question by showing that there is no justifiable basis for privileging the molecular level. I arrived at the germs of a Theory of Biological Relativity[9] from a strongly reductionist position. So I know the strengths and attraction of reductionism. But my early research forced me to rethink.

When I was working as a research student at University College London, I studied ion channels in the heart, as described in the Interview. At that time such work was about as reductionist as you could get in physiology. It was also a field of physiology to which mathematics could be readily applied with great success. I was able to construct the first useful mathematical model of heart rhythm. But it was precisely the use of mathematics that led to my abandoning the hard reductionist viewpoint. I was forced to admit that reductionism had to be softened by respect for multilevel causation.

The way in which that happened depended on an insight of Alan Hodgkin. When solving the differential equations that he developed with Andrew Huxley to study the nerve impulse, he realized that those equations necessarily require downward causation from the level of the whole cell, as well as upward causation from the level of molecules. This is what we now call the "Hodgkin Cycle." My work showed that the same cycle must operate in the heart, and indeed in all electrically excitable cells.

Now, you might think that this is just a feature of excitable cells. In fact, it is a feature of all biological processes that can be described

---

[9] The sense in which the statement that "there is no privileged level of causation" is a relativity statement is that it follows the general principle of relativity, which is to distance ourselves from privileging frames of reference for which there is no justification. Einstein's theories of relativity also follow this general principle. For more details on this issue, see D. Noble, *Dance to the Tune of Life: Biological Relativity* (Cambridge UP, 2017).

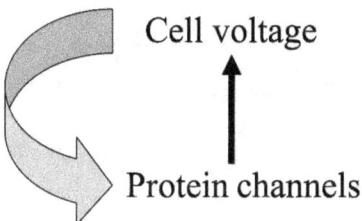

Figure 4.3: *The Hodgkin Cycle. Cell voltage (a global property of the whole cell) controls the protein ion channels, which at the same time conduct ionic current that changes the cell voltage. This is circular causality between the two levels.*

by differential equation models. It is hard to think of any biological processes that could not be described by such equations since processes (i.e., how components change with time or space or any other dimension that allows derivatives to quantify those changes) are precisely what differential equation models are designed for.

Now I come to a key point. Even a Laplacian determinist would have to admit the existence of this contextual dependence of biological processes described by such models. That point was realized by the philosopher Benedict de Spinoza in 1665. I was introduced to Spinoza's work by Stuart Hampshire—the philosopher whose seminars I gate-crashed at University College London nearly 60 years ago. Much later I found that, well before Laplace, Spinoza correctly described such contextual dependence of biological processes. The evidence is in a letter that the great Dutch-Jewish philosopher Baruch (Benedict) Spinoza wrote in Latin to Henry Oldenburg, the first secretary of the Royal Society, which is still kept in the Royal Society archives.[11]

---

[11] See B. Spinoza, *The Letters,* tr. S. Shirley. Indianapolis: Hackett Publishing Co., 1995; Letter #32, pp. 192–198. (Note that the text cited is from an earlier translation of the *Letters* by R.H.M. Elwes.—Ed.)

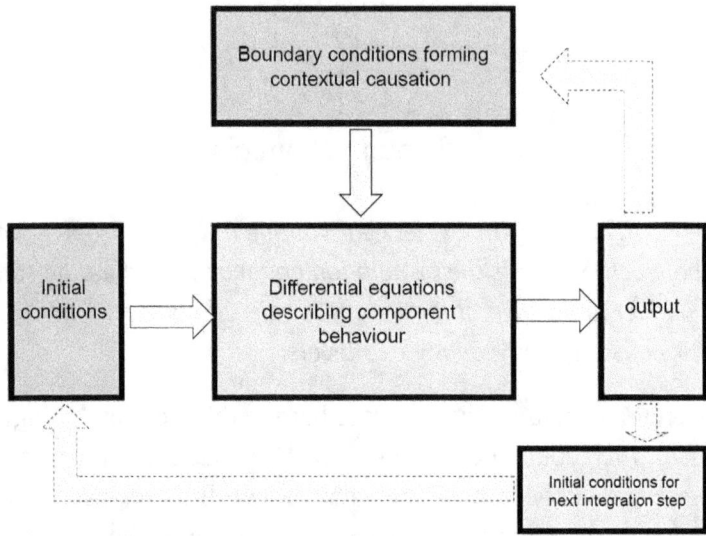

Figure 4.4: *Many models of biological systems consist of differential equations for the kinetics of each component. These equations cannot give a solution (the output) without setting the initial conditions (the state of the components at the time at which the simulation begins) and the boundary conditions. The boundary conditions define what constraints are imposed on the system by its environment and can therefore be considered as a form of contextual causation from a higher scale. The arrows are not really unidirectional. The dotted arrows complete the diagram to show that the output contributes to the boundary conditions (although not uniquely), and they determine the initial conditions for the next integration step.*[10]

---

[10] The legend and diagram are from D. Noble, "A Theory of Biological Relativity: No Privileged Level of Causation," *Interface Focus*, 2012, **2**: 55–64. This diagram is highly simplified to represent what we actually solve mathematically. In reality, boundary conditions are also involved in determining initial conditions and the output parameters can also influence the boundary conditions, while they in turn are also the initial conditions for a further period of integration of the equations. There are also important differences between ordinary differential equation models and partial differential equation models. The boundary conditions in partial differential equations become incorporated into the parameters in equivalent ordinary differential equation models.

The Latin text of the section translated into English here begins *"concipiamus jam, si placet, ..."* (Let us imagine, with your permission, ...). The full English translation of the relevant section is:

> Let us imagine, with your permission, a little worm, living in the blood, able to distinguish by sight the particles of blood, lymph, etc., and to reflect on the manner in which each particle, on meeting with another particle, either is repulsed, or communicates a portion of its own motion. This little worm would live in the blood, in the same way as we live in a part of the universe, and would consider each particle of blood, not as a part, but as a whole. He would be unable to determine, how all the parts are modified by the general nature of blood, and are compelled by it to adapt themselves, so as to stand in a fixed relation to one another.

This paragraph could stand even today as a succinct statement of one of the main ideas of Biological Relativity. He doesn't use a mathematical format to express his idea (this was before the development of Newton's mechanics), but as I have shown, the idea is beautifully expressed by the mathematics of differential equations. The reason is that, without the initial and boundary conditions, such equations cannot be solved.

Now, Spinoza was a strict determinist, so strict that he did not even distinguish reasons from causes. Everything in the universe envisaged by Spinoza was part of a giant piece of clockwork. The Theory of Biological Relativity therefore requires two more facts to be added. The first is that the universe is not a massive piece of clockwork. Stochasticity rules everywhere, and it most certainly rules in biological organisms. This is true, for example, of all forms of expression of individual proteins. If you take a cultured cell population and measure the expression level of any protein, you find that it can vary between 10- and 1000-fold between different cells in the population.

Not only is the stochasticity extensive, it is itself controlled by the cell population as a whole. If a new population of cells is cultured from outliers in the original population, the new population

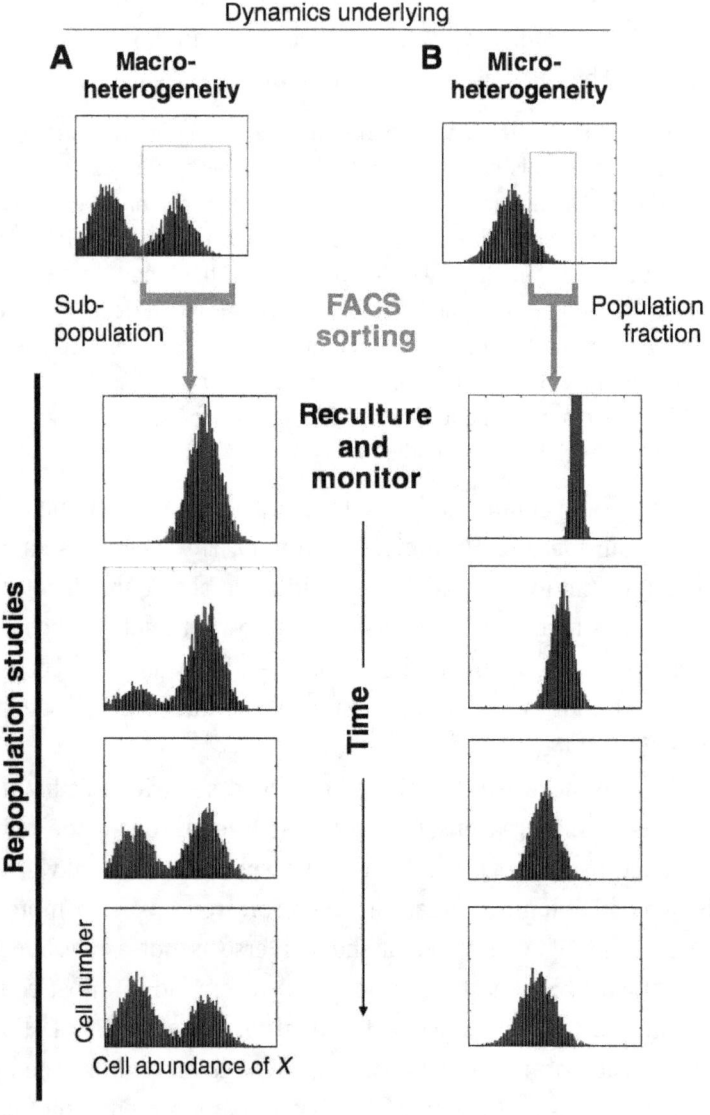

Figure 4.5: *Diagram illustrating two forms of stochasticity in protein expression in cell populations, bimodal (left) and monomodal (right). Experiments show that if you select cells from one of the two distributions, or from outliers in a mono-modal set, the new population initially expresses the distribution of the mother cells. But after a few days the whole population reverts to the original distribution. Thus, the population (not individual cells) determines the distribution. From the work of Sui Huang.*[12]

reverts after a few days to the distribution displayed by the original population. This is therefore an attractor; moreover, the attractor is a property of the whole population, not of individual cells. This is yet another example of the contextual dependence of biological systems.

Huang's experiments show the control of stochasticity in cultured cell populations. Does this control of stochasticity also occur across generations of intact organisms? Surprisingly, perhaps, this was demonstrated by the great Danish botanist Wilhelm Johannsen as long ago as 1911. The images below come from the lecture by Michael Joyner referred to above.

If cell populations can manipulate stochasticity to this degree, then they can also use it functionally. I will give just one example of that, and it is one that is highly relevant to evolutionary biology since it concerns the way in which organisms can change their own genomes in a directed, functional way.

The example is from the immune system. The germ line has only a finite amount of DNA. In order to react to many different

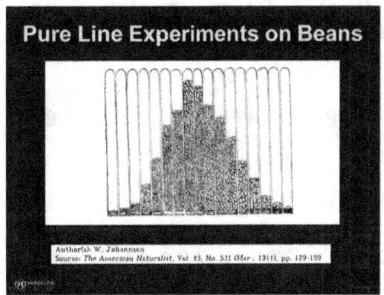 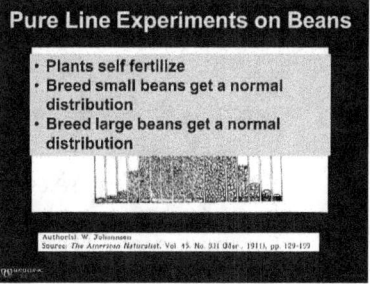

Figure 4.6: *Images from Joyner lecture, showing Johannsen's 1911 experiment on beans. The same normal distribution occurs whether one breeds from small or large bean plants.*[13]

---

[12] S. Huang, "Non-genetic Heterogeneity of Cells in Development: More Than Just Noise," *Development,* 2009, **136**: 3853–3862; H.H. Chang, M. Hemberg, M. Barahona, D.E. Ingber, and S. Huang, "Transcriptome-wide Noise Controls Lineage Choice in Mammalian Progenitor Cells," *Nature,* 2008, **453**: 544–548.

[13] W. Johannsen, "The Genotype Conception of Heredity," *American Naturalist,* 1911, **45**: 129–159.

antigens, lymphocytes "evolve" quickly to generate extensive anti-gen-binding variability. There can be as many as 1012 different antibody specificities in the mammalian immune system, and the detailed mechanisms for achieving this have been known for many years. The mechanism is directed, because the binding of the antigen to the antibody itself activates the proliferation process. The antigen activates special lymphocytes (cells in the blood stream) called B-cells, which evolve rapidly to generate a huge range of antigen-binding variability.[14] Targeted speeding-up of change is therefore one mechanism by which functional change can occur. That is true even if the individual changes at that location are random. The functionality lies in the targeting of the location. That targeting is not random.

This example should warn us that the simple idea that random changes necessarily exclude functional changes is incorrect. Organisms can use random changes to exploit their world and develop their own inheritance in a functional way. The example from the immune system concerns inheritance of cells within a single generation of a multicellular organism. Later, I will give examples of such inheritance across generations. Yet strict neo-Darwinism excludes these processes. This is how it excludes Lamarckian (i.e., functionally directed) forms of inheritance.

So far, I have outlined two essential features leading to the Theory of Biological Relativity: (1) downward as well as upward causation, meaning that molecules are constrained by their context in biological systems, and (2) stochasticity, which is itself under control by the organism.

There is a third requirement for the full theory, which is (3) the creation of novelty, often at new levels of selection. A multi-level theory of organisms could not exist, after all, if evolution did not create novel levels of activity and selection. Evolution has done that many times, for example in the creation of cells, the formation of

---

[14] For further details, see J.A. Shapiro, *Evolution: A View from the 21st Century*, Upper Saddle River, NJ: FT Press, 2011; pp 66–68.

eukaryotes, and many other forms of symbiosis and symbiogenesis, and the emergence of population levels of selection. The creation of novelty itself then changes the dynamics of evolution since organisms actively change their environment. In all these processes the equations often favored in evolutionary biology, which involve equilibrium dynamics and similar agents, will not capture what is happening. The dynamic creation of novelty leads to dynamic changes in the environment and the creation of new niches.

When nature tells us that our mathematical models are too restrictive, we should listen carefully. We now need ideas and models that respect what nature is telling us, loud and clear, about the imitations of our existing models.

Full details on the Theory of Biological Relativity will appear in my most recent book, *Dance to the Tune of Life: Biological Relativity* (Cambridge UP, 2017). The account given here is necessarily brief.

4.3. The Gene-Centric View Has Damaging Consequences

The gene-centric view has had profoundly damaging (even if not intended) consequences in sociology, economics, politics, and many other areas of the humanities and social sciences.

I am not a social historian, so I tread here very lightly. My impression is that Social Darwinism, the eugenics movement, and many evolution-based theories of economics, finance, and politics were greatly influenced by the standard gene-centric theory of evolutionary biology that dominated the twentieth century. In economics and management, that is still true.[15] It is also clear that some of these social developments were disastrous.

In noting this, I am not of course apportioning any blame. Nor am I by any means the only biologist to have this impression. Conrad

---

[15] See, for example, *Journal of Economic Behavior & Organisation*, volume **53**, 2004, which is devoted to articles in this cross-disciplinary area.

Waddington wrote way back in 1957 that:[16]

> Many humanist and religious authors... have drawn attention to its
> [Neo-Darwinism's] damaging effects on man's spiritual life.

He specifically referred to non-religious, as well as religious, authors,
so he was clearly not using the word "spiritual" in a purely religious
context. On the contrary, he viewed ethics and other aspects of spiri-
tuality as a secular matter in the context of evolutionary progress.[17]

Moreover, leading neo-Darwinians have also acknowledged the
danger, explicitly or implicitly. Consider, for example, this quotation
from Richard Dawkins:[18]

> Let us try to teach generosity and altruism, because we are born
> selfish. Let us understand what our own selfish genes are up to,
> because we may then at least have the chance to upset their designs,
> something that no other species has ever aspired to.

This is a fairly explicit acknowledgement of the need to "upset their
designs." I will leave to a later section whether selfish-gene theory
really can mean that "we are born selfish."

4.4. The Gene-Centric View Resists New Findings

The gene-centric view has had to gyrate in a contorted way to accommodate one new finding after another. The final straw for me was a
supporter of neo-Darwinism purporting to accept the inheritance of
acquired characteristics. This is like eating your own tail.

The gene-centric view has had to gyrate in a contorted way to
accommodate one new finding after another. The final straw for me

---

[16] C.H. Waddington, *The Strategy of the Genes*. London: George Allen & Unwin, 1957.

[17] See Peter J. Bowler, *Reconciling Science and Religion*. Chicago: University of
Chicago Press, 2001; p. 80: "Waddington had no interest in encouraging scientists to revive
an interest in religion." This is also clear from Waddington's 1942 book, *Science and Ethics*.
His concern was for scientists to be involved in the social process. To quote Bowler again:
"...his approach to ethics was more in line with Huxley's vision of humanity continuing the
course of evolutionary progress by more efficient means."

[18] R. Dawkins, *The Selfish Gene*. Oxford: Oxford University Press, 1976; p. 3.

was a supporter of neo-Darwinism purporting to accept the inheritance of acquired characteristics. This is like eating your own tail.

First, we need to ask what we are talking about. Neo-Darwinism is significantly different from Darwinism, in the sense of Charles Darwin's own thought, although many neo-Darwinists give the impression that they can be conflated. In his On the Origin of Species (1859) and in his later books, Darwin presented a much more nuanced position. Natural Selection was argued to be the most important— but not the only—mechanism of evolution.[19]

Specifically, Darwin, like Lamarck before him, accepted the idea of the inheritance of acquired characteristics. Ernst Mayr notes about 12 places in *On the Origin of Species* where Darwin does this.[20] Even more significant, in *The Variation of Animals and Plants under Domestication* (1868), Darwin takes the idea so seriously that he formulated a theory for how it might happen. He realized of course that in multicellular organisms with separate germ lines it would be necessary for some information about changes in the organism to be transmitted to the germ line for it to be possible for acquired characteristics to be inherited. His tentative solution was his theory of "gemmules," small particles travelling in the blood to transmit this information. We now know that small RNAs can play just such a role. I will return to that comparison later.

By contrast, neo-Darwinism can be seen, historically at least, as specifically excluding this process. For its founder, August Weismann, this was an absolutely central feature of the theory. He wrote:[21]

---

[19] Darwin wrote in the first edition of *On the Origin of Species*: "I am convinced that natural selection has been the main, but not the exclusive means of modification," a statement he reiterated with increased force in the sixth edition (1872).

[20] E. Mayr, "Introduction," in C. Darwin, *On the Origin of Species*. Cambridge, MA: Harvard University Press, 1964; pp. xxv–xxvi. Mayr writes: "Curiously few evolutionists have noted that, in addition to natural selection, Darwin admits use and disuse as an important evolutionary mechanism. In this he is perfectly clear."

[21] A. Weismann, *Das Keimplasma; eine Theorie der Vererbung*. Jena: Fischer, 1892. (Published in English as *The Germ-Plasm: A Theory of Heredity*. New York: Scribner, 1893).

When these deviations only affect the soma, they give rise to temporary non-hereditary variations; but when they occur in the germ-plasm, they are transmitted to the next generation and cause corresponding hereditary variations in the body.

Regarding the sufficiency of the genome, Ernst Mayr wrote:[22]

All of the directions, controls and constraints of the developmental machinery are laid down in the blueprint of the DNA genotype as instructions or potentialities.

Richard Dawkins summed it up when he wrote that genes are "sealed off from the outside world."[23]

But, if we accept that environmentally induced phenotypes can be inherited, as recent observations show and which I will discuss in a later section, then we have broken the Weismann barrier, because the germline is no longer isolated from the environment and the organism's response to it. We have also automatically broken the other neo-Darwinian assumption of random variation because phenotype changes can then guide inheritable variation, at least to some degree, as we have seen in item 2, above. The honest response to this situation is to say that the central tenets of neo-Darwinism are simply no longer valid.[24] We then return to a modern version of Darwinism, in the sense of Darwin's original thinking. The reason for distinguishing neo-Darwinism from Darwinism disappears.

The other key feature of neo-Darwinism was that all genetic change is random with respect to function, which would exclude any other way in which acquired characteristics could be inherited. This assumption was also first introduced by August Weismann. It is worth noting though that some neo-Darwinists have questioned the basis of

---

[22] E. Mayr, "The Triumph of the Evolutionary Synthesis," *Times Literary Supplement*, November 2, 1984; p. 126.

[23] R. Dawkins, *The Selfish Gene*. Oxford: Oxford University Press, 1976; p. 21.

[24] D. Noble, "Central Tenets of neo-Darwinism Broken. Response to 'Neo-Darwinism is Just Fine,'" *Journal of Experimental Biology*, 2015, **218**: 2659.

that assumption. An example is John Maynard Smith's statement in his book *Evolutionary Genetics*:[25]

> ...it is not clear why he thought it [Weismann's claim that the germ line is independent of the soma] was true.

Neo-Darwinism therefore led to an unjustified narrowing-down of the mechanisms supposed to have been involved in the origin of different species. Moreover, the motivation for the development of neo-Darwinism was clearly a move to exclude Lamarckism. Despite Darwin's acceptance of the idea, the inheritance of acquired characteristics was deemed impossible. Natural selection working on chance variations in genetic material was thought to be entirely sufficient to explain all evolutionary change. The experimental evidence for this was flimsy, as John Maynard Smith acknowledged.

I believe that it was this unnecessary insistence on the two central assumptions of Weismann, combined with the apparent strength of support from molecular biology and the Central Dogma, that created the conditions that required some contorted gyrations to accommodate all the findings that could be regarded as challenging either of the basic assumptions. The consequence though was unfortunate since some very important discoveries were side-lined or reinterpreted in ways that closed off lines of research that should have been followed up. I will briefly list here some of the examples.

Spalding, Baldwin, and the adaptability driver. This is a phenomenon usually known as the "Baldwin effect,"[26] or "adaptability

---

[25] J. Maynard Smith, *Evolutionary Genetics*. Oxford: Oxford University Press, 1998; p 9.

[26] J.M. Baldwin, "A New Factor in Evolution," *American Naturalist,* 1896, **30**: 441–451. Many biologists today regard the Baldwin effect as perfectly compatible with the neo-Darwinian Modern Synthesis. This can be done by taking a purely gene-centric view of what is happening. That viewpoint conceals the fact that the process depends on an active choice of environment at the level of the phenotype. Baldwin was a psychologist and described the phenomenon as the effect of learned behavior on evolution. The important point is that it is organisms that choose to behave in a particular way, not genes. The active role here occurs at the phenotype level. Genes then follow by the process of assimilation. The Baldwin effect is as much an assimilation of a character into the gene pool as Waddington's experiments were.

driver"—which is the term I prefer.[27] Organisms can choose new niches for themselves and their descendants.[28] Moving to a new niche can change the course of evolution even with no mutations whatsoever. That choice is a physiological characteristic of the phenotype, not a change in DNA. So how can it change the course of evolution? The answer is surprisingly simple. In a wild population in which individual genomes are not identical, the combinations of alleles in the adventurous organisms discovering new niches will be favored. That is an evolution of the genome by combinatorial selection, not selection of new random mutations. Such selection can lead to inherited novelty, as Waddington showed so clearly in his work on fruit flies.

Conrad Waddington and genetic assimilation. The Weismann Barrier was based on some experiments in which the amputation of tails in mice did not lead to mice being born with no tails. But surgical mutilation is not a test for Lamarckian forms of inheritance. Waddington performed the experiments that more successfully tested for the inheritance of acquired characteristics by using environmental manipulations that played into natural plasticity in fruit fly populations. But orthodox neo-Darwinists dismissed Waddington's findings as merely an example of the evolution of phenotype plasticity. That is what you will find in many of the biology textbooks. I think that is to misrepresent what Waddington showed. Of course, plasticity can evolve, and that itself could be by a neo-Darwinist or Darwinist or any other mechanism. But Waddington was not simply showing the evolution of plasticity in general; he was showing how it could be

---

[27] This is the term favored by Patrick Bateson, who has carefully researched the literature on the "Baldwin Effect:" see P. Bateson, "The Adaptability Driver: Links between Behavior and Evolution," *Biological Theory*, 2006, 1: 342–345. He makes two points. First, that this process was first identified by Douglas Spalding in 1873, so predating Baldwin. Second, that a behavioral driver dependent on the adaptability of the phenotype could drive evolution more rapidly in a direction that would be extremely unlikely to occur by combinations of chance mutations. Both of these conclusions are convincing.

[28] The mechanisms by which animals make such choices (sometimes known as "niche construction") is now an active field of study; see, for example, the work of Daniel Rubenstein.

exploited to enable a particular acquired characteristic in response to an environmental change to be inherited and become assimilated into the genome.

Barbara McClintock and "jumping genes." Barbara McClintock won a Nobel Prize in 1983 for her discovery in the 1940s of mobile genetic elements, often called "jumping genes." She correctly saw that this meant that the genetic material is under some form of control by the organism, i.e., that—as she put it:[29]

the genome... [is] a highly sensitive organ of the cell.

Yet, by 1957 McClintock was completely discouraged from publishing further work on the subject—until she received the Nobel Prize in 1983 at the age of 81.

As James Shapiro has shown in his book *Evolution: A View from the 21st Century* (FT Press, 2011), large-scale re-organization of genomes has occurred during evolution. These changes are sometimes represented in standard theory as "large mutations." The problem with that designation is that, unlike the gradual accumulation of point mutations, the domains that shift around can carry functionality with them.

Carl Woese and the discovery of archaea. Carl Woese was described by Science in 1997 as "microbiology's scarred revolutionary" because his discovery of *archaea* as a separate domain[30] was rubbished by neo-Darwinists like Ernst Mayr.[31] Another great contribution to the study of the prokaryotes was also made by Carl Woese, which is that archaea share the bacterial propensity for promiscuous sharing of DNA. Horizontal gene transfer has occurred and still occurs frequently amongst prokaryotes, and also occurs to some degree

---

[29] B. McClintock, "The Significance of Responses of the Genome to Challenge," *Science*, 1984, **226**: 792–801; p. 800.

[30] C.R. Woese and G.E. Fox, "Phylogenetic Structure of the Prokaryotic Domain: The Primary Kingdoms," *Proceedings of the National Academy of Sciences, USA*, 1977, **74**: 5088–5090; doi: 10.1073/pnas.74.11.5088.

[31] E. Mayr, "Two Empires or Three?," *Proceedings of the National Academy of Sciences, USA*, 1998, **95**: 9720–9723

amongst eukaryotes. The whole field of microbiology is a problem for neo-Darwinism, since there is no separate germ line, and there is rampant exchange of DNA between species. Evolution is as much dependent on horizontal transfer of DNA as on vertical inheritance. Even the concept of species is a problem.

Lynn Margulis and symbiogenesis. I will discuss this example in the next section.

What precisely has been disproved by these and other developments that challenge neo-Darwinism? This is an important question.

First, I want to note that the standard neo-Darwinist mechanism, i.e., the accumulation of gradual mutations followed by natural selection, has not been disproved. What has been disproved is the idea that this is the *only* way in which evolutionary change can happen. For example, the equations of population genetics, which are based on neo-Darwinist mechanisms, could still be valid as descriptions of the particular conditions and processes they were designed to describe, though we should note that those conditions are highly idealized. A rough analogy is the way in which Newtonian mechanics has been replaced by quantum mechanics. That does not invalidate the use of Newton's equations in many situations to which they apply well enough. But, as with the move in physics to acknowledge what nature tells us by developing new models, we should move on to new mathematics when it becomes clear that our existing maths is inadequate for the job.

This last point is important because some evolutionary biologists seem to fear replacement of neo-Darwinism by a more nuanced, multi-process view of evolution. Some of the arguments about whether neo-Darwinism has been extended or replaced seem to me to be largely a matter of viewpoint. I find myself more comfortable with a replacement view. Others feel more comfortable with an extension view. It is as simple as that. Even those who favor a replacement view can acknowledge, as I do, the continued existence of ranges of application for which the neo-Darwinist mechanism is valid.

4.5. The Gene-Centric View Claims Parsimony

Nature simply isn't parsimonious.

The appeal to parsimony was inherent in Weismann's reactions to the nineteenth-century neo-Lamarckians. His 1893 response to Herbert Spencer, for example, was entitled *Die Allmacht der Naturzüchtung* [The All-sufficiency of Natural Selection]. He wrote:[32]

> We accept it [Allmacht] ... simply because we must, because it is the only plausible explanation that we can conceive.

He admitted that it was not possible to observe the process in detail, so there could be no experimental proof, but continued:

> It does not matter whether I am able to do so or not, or whether I could do it well or ill; once it is established that natural selection is the only principle which has to be considered, it necessarily follows that the facts can be correctly explained by natural selection.

He doesn't fully explain what is meant by "the only principle which has to be considered," but he does admit that it doesn't depend on any experimental proof. It was seen as just "necessary."

We encounter a similar approach in a 2009 debate between Richard Dawkins and Lynn Margulis.[33] In this debate, the following exchange occurred:

> **Dawkins**: It [neo-Darwinism] is highly plausible, it's economical, it's parsimonious, why on earth would you want to drag in symbiogenesis

---

[32] A. Weismann, *Die Allmacht der Naturzüchtung; eine Erwiderung an Herbert Spencer.* Jena: Fischer, 1893. (This "rejoinder" to Herbert Spencer's criticisms of Darwin's theory sparked a lively exchange in the pages of the Contemporary Review: see Herbert Spencer, "The Inadequacy of Natural Selection: Professor Weismann's Theories: A Rejoinder to Professor Weismann," *Contemporary Review*, 1893, vol. **64** [Feb., Mar., May, and Dec.]; and A. Weismann, "The All-Sufficiency of Natural Selection," *Contemporary Review*, 1893, vol. **64** [Aug. and Sep.]. The passages cited in the main text of this Statement are taken from Weissman's contribution to the *Contemporary Review* exchange [p. 336], and are quoted by S.J. Gould, *The Structure of Evolutionary Theory*, Cambridge, MA: Harvard University Press, 2002; p. 202—Ed.)

[33] [https://www.youtube.com/watch?v=YJ-sZHHx7O0.]

when it's such an unparsimonious, uneconomical [theory]?

Margulis: Because it's there.

That's it in a nutshell. What is there, what exists, is the starting point of all science.

Symbiogenesis is indeed more complex, but nature has clearly used it many times over.

In 2014, I and my colleagues wrote an editorial in an issue of the Journal of Physiology devoted to evolutionary biology. We concluded:[34]

> Nature is even more wondrous than the architects of the Modern Synthesis thought, and involves processes we thought were impossible.

We are still absorbing the immense implications of these developments in biology, and many other disciplines. Whole areas of economics, sociology, and philosophy are based on interpretations of selfish gene viewpoints. No field of human endeavor will remain untouched since the implications affect even our concept of humanity.

4.6. The Gene-Centric View Claims to Settle the Question of Lamarckism

The claim that the Weismann Barrier and the Central Dogma have settled the question whether Lamarckism is possible. But Weismann's experiments were not a test for Lamarckism and the Central Dogma does not counter the fact that the organism controls the genome.

August Weismann performed his tail amputation experiments in 1890. They were designed to counter rather wild claims of nineteenth-century neo-Lamarckians who thought that surgical changes

---

[34] D. Noble, E. Jablonka, M.J. Joyner, G.B. Müller, and S.W. Omholt, "Evolution Evolves: Physiology Returns to Centre Stage," *Journal of Physiology*, 2014, **592**: 2237–2244.

could be inherited. Some even claimed that the practice of circumcision could reduce or eliminate the foreskin in the offspring.

But this was not Lamarck's idea. His idea was that inheritance may occur in a functional interaction between the organisms *and their environment,* through use and disuse of the organism's structures. The question is not whether the *non-functional results of surgery can be inherited.* Darwin must have known already that such inheritance did not occur from the work of animal breeders. Tail amputation in dogs for aesthetic reasons does not result in stunted tails in the offspring, no matter how many generations are bred from the animals.

The real question—to put it in a more modern form—is whether the germ-line is or is not isolated from environmental influences. The relevant way to do a tail-cutting experiment or any other experiment to answer *that* question would be to change the environment in a way that makes taillessness a functional advantage. Quite apart from the obvious question why a surgical change should be inherited, even a standard Lamarckian would notice that the environment, apart from the surgery, is not different. Furthermore, even if there were environments that would favor taillessness, the experiment would not test for that. Weismann's test for Lamarckism simply would not pass the elementary tests for a scientific experiment today.

The work of Conrad H. Waddington showed the more successful way forward for such experiments. The way to test for the inheritance of acquired characteristics is first to discover what forms of developmental plasticity already exist in a population, or which the population could be persuaded to demonstrate with a little nudging from the environment. This approach is more finely nuanced than using surgery since it is playing into plasticity that is already present.

Waddington used the word "canalized" for this kind of persuasion since he represented the developmental process as a series of "decisions" that could be represented as "valleys" and "forks" in a developmental landscape. He knew from his developmental studies that embryo fruit flies could be persuaded to show different wing structure simply by changing the environmental temperature or by a chemical stimulus. In the developmental landscape this could be represented as a small manipulation in slope that would lead to favoring one channel in the landscape rather than another, so that the adult could show a different phenotype starting from the same genotype.

The next step in his experiment was to select for and breed from the animals that displayed the new characteristic. Exposed to the same environmental stimulus, these gave rise to progeny with an even higher proportion of adults displaying the new character. After around 14 generations Waddington found that he could then breed from the animals and obtain robust inheritance of the new phenotype characteristics even without applying the environmental stimulus.[35] The characteristics had therefore become locked into the genetics of the animal. He called this process "genetic assimilation." Since the plasticity being exploited already existed, it is likely that what happened is that all the gene variants (alleles) for the characteristic already existed in the population, so that selection could bring that set of variants together in an individual.

Waddington's work was largely sidelined by most evolutionary biologists. It did not become part of the mainstream. So, an opportunity to develop a more inclusive, systems, and multi-scale approach to evolutionary biology was lost. A plausible explanation is that this was the period when molecular biology was rapidly developing and had already become dominant. Francis Crick's so-called "Central Dogma" of molecular biology was published in 1956,[36] just before Waddington published his book *The Strategy of the Genes* (George Allen & Unwin, 1957). There is no doubt which has had the largest and longest impact. Sadly, Waddington was largely forgotten. So, also, were many lessons from integrative physiology.

---

[35] C.H. Waddington, "The Genetic Assimilation of the Bithorax Phenotype," *Evolution*, 1956, **10**: 1–13; see, also, D. Noble, "Conrad Waddington and the Origin of Epigenetics," *Journal of Experimental Biology*, 2015, **218**: 816–818.

[36] F. Crick,"On Protein Synthesis," *Symposia of the Society for Experimental Biology*, 1956, **12**: 139–163.

Part of the reason lies in the way in which Crick's Central Dogma was greeted by neo-Darwinists as welcome and impressive support for the Weismann Barrier idea. The two reinforced each other. The isolation of the germ-line seemed to be confirmed spectacularly by the finding that DNA codes for proteins through the intermediate of RNA, whereas protein sequences do not code for DNA or RNA. This is represented by the shaded downward-pointing arrows below.

The actual situation is much more complex and has been widely misunderstood.

First, the 1956 version of the Central Dogma had to be substantially revised in 1970 when it was discovered that the step from DNA to RNA is in fact reversible (upward white arrow). This is one of the processes that enable whole domains of DNA sequence

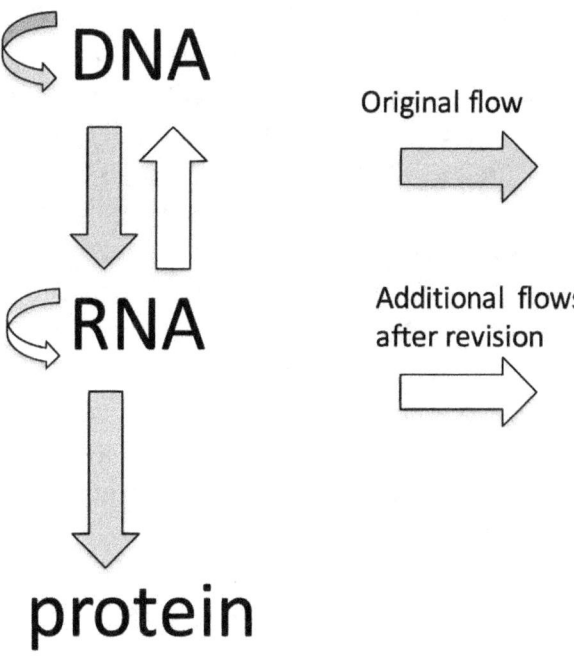

Figure 4.7: *The so-called "Central Dogma" of molecular biology, original and revised versions.*

to move around the genome, with very important consequences in evolutionary history.

Second, the way in which Crick formulated the Dogma makes it clear that it actually refers only to transmission of sequence information. That is only one half of the story on the interactions between an organism, its genome, and the environment. Crick wrote:[37]

> The central dogma of molecular biology deals with the detailed residue-by-residue transfer of sequential information. It states that *such information* cannot be transferred back *from protein* to either protein or nucleic acid.

I have italicized "such information" and "from protein" since it is evident that the statement does not say that no information can pass from the *organism* to the genome. Crick must have known that absolute isolation of the genome from control information could not be true. How else could the same genome be used by the many different cells, tissue, and organs of the body to generate very different phenotypes? Note also that the statement refers to transfer back *from proteins*. The information that regulates gene expression via transcription factors and epigenetic marks comes, of course, from the networks as a whole, not from individual proteins, although the final message is conveyed by proteins called "transcription factors." It is the *pattern* of such factors that is important and that it is a global property of the cells, tissues, and organs involved. There are many possible patterns of transcription factors, each of which corresponds to a different phenotype outcome. The information that passes from the system to the genome is of a different kind from that involved in coding. It is not a property of individual molecular sequences, but rather a property of an ensemble.

The more we learn about DNA and the chromosomes, the more we find that they are more like a database used by the rest of the organism, and that Barbara McClintock was right when she wrote that the genome is an "organ of the cell." We have come a long

---

[37] F. Crick, "The Central Dogma of Molecular Biology," *Nature*, 1970, **227**: 561–563.

way since Descartes fired the first shot in the reductionist-mechanist agenda with this statement in his treatise on the fetus:[38]

> If one had a proper knowledge of all the parts of the semen of some species of animal in particular, for example of man, one might be able to deduce the whole form and configuration of each of its members from this alone, by means of entirely mathematical and certain arguments, the complete figure and the conformation of its members.

I wonder what Descartes would think of the modern experiments on cross-species clones. If Descartes, Weismann, and Crick were right, then transferring the nucleus of one species into the fertilized egg cell of another species to replace its removed nucleus should unambiguously lead to an organism that matches the blueprint of the nucleus. So, what do we find? First, they would be shocked to find that for most cross-species clones, the experiment doesn't even work. Usually, the embryonic development freezes at some point. There is therefore an incompatibility between the genetic material of the donor nucleus and the recipient egg cell. Second, in the rare cases where the experiment works, we obtain an organism intermediate between the two species.

The most spectacular example of this kind of experiment comes from work done at the Wuhan Fish Institute in China by Yonghua Sun and his colleagues in 2005 using two different species of fish, where the nucleus of one species was used to replace the nucleus in a fertilized egg cell of the other species. The outcome in the anatomy of the adult that resulted from this cross was determined by the cytoplasmic structures and expression patterns of the egg cells, as well as the transferred DNA. The basic features of structural organization both

---

[38] R. Descartes, *La formation du foetus* (Paris, 1664). The original French text reads: "*Si on connoissoit quelles sont toutes les parties de la semence de quelque espece d'Animal en particulier, par exemple de l'homme, on pourroit déduire de la seul, par des raisons entierement Mathematiques et certaines, toute la figure & conformation de ses membres...*" (paragraph LXVI; p. 146). (Note that *La formation du foetus* was first published posthumously, bound together in one volume with *L'homme*. We have respected the seventeenth-century spelling and punctuation of the original. The English translation is by Denis Noble and Anthony Kenny and has not been previously published—Ed.)

Figure 4.8: *Cross-species clone. The nucleus of a common carp,* Cyprinus carpio *(middle), was transferred into the enucleated egg cell of a goldfish,* Carassius auratus *(left). The result is a cross-species clone (right) with a vertebral number closer to that of a goldfish (26–28) than of a carp (33–36) and with a more rounded body than a carp. The bottom illustrations are X-ray images of the animals in the top illustration. Figure kindly provided by Professor Yonghua Sun.*[39]

of cells and of multicellular organisms must have been determined by physical constraints before the relevant genomic information was developed.

4.7. The Gene-Centric View Claims That Epigenetic Inheritance is Short-Lived

The claim that epigenetic inheritance always dies out after a generation or two. There are clear examples where it doesn't, and in any case no one supposes that an evolutionary change initiated by epigenetic effects would be the consequence of a single-generation exposure

---

[39] Y.H. Sun, S.P. Chen, Y.P. Wang, W. Hu, and Z.Y. Zhu, "Cytoplasmic Impact on Cross-Genus Cloned Fish Derived from Transgenic Common Carp (*Cyprinus carpio*) Nuclei and Goldfish (*Carassius auratus*) Enucleated Eggs," *Biology of Reproduction*, 2005, **72**: 510–515.

to the changed environment. Multiple-generation exposures can be assimilated into the genome.

Epigenetics includes a variety of ways in which the genome is controlled by the organism. The word "epigenetics" was introduced by Waddington to refer to what he called the "canalization" of development leading to different possible outcomes using the same genome. There are many different cell types in multicellular organisms. They differ in the expression patterns of their proteins. Waddington clearly saw that the complex cell networks are responsible for enabling that to happen. That should have been a signal to biologists that studying those networks was just as important as studying genes.

More recently, the concept of epigenetics has been greatly extended to include processes by which DNA itself and the chromatin proteins can be marked by chemicals that alter the probability of gene expression. The existence of these marking processes has made it easier to understand how the genome is controlled to become "an organ of the cell." Inevitably, this development has also raised an important question in evolutionary biology: Since these marks are inherited within cell populations in a single generation, can they also be inherited across generations?

The initial reaction of neo-Darwinists was to propose that any such inheritance of genome marking was either impossible or, if it did happen, it would be found to be only transitory, dying out after a generation or two.

Before I come to the question whether this is always true and whether such marking can become more permanent, I want to acknowledge an important respect in which transient DNA marking could be very important in evolution. The ability of a population to experiment with transient changes would enable it to explore options in a reversible way that may be critical to survival in times of environmental stress. Warren Burggren has recently reviewed this question and concludes:[40]

> ...when environments are dynamic (e.g., climate change effects), there may be an "epigenetic advantage" to phenotypic switching by epigenetic inheritance, rather than by gene mutation. An epigeneti-cally-inherited trait can arise simultaneously in many individuals, as opposed to a single individual with a gene mutation. Moreover, a transient epigenetically-modified phenotype can be quickly "sun-setted," with individuals reverting to the original phenotype. Thus, epigenetic phenotype switching is dynamic and temporary and can help bridge periods of environmental stress. Epigenetic inheritance likely contributes to evolution both directly and indirectly.

This could work the other way, also. If the environment change is more long-term, there need not be a switch-back. The process of genetic assimilation could then make the change permanent, just as happened in Waddington's experiments.

Evidence that this may have happened in evolutionary history has come from the study of one of the icons of Darwinian evolution: the finches of the Galápagos Islands. Michael Skinner and his team have investigated both the genetic and epigenetic changes that have occurred in these finches. The answer is that both have occurred and that the epigenetic changes correlate rather better with phylogenetic distance than do the genetic changes. The evolution of these species has almost certainly involved both changes and interactions between them.

One of the criticisms neo-Darwinists raise when the role of epigenetics in transgenerational change is proposed is that such changes always die out. That may well be true in most cases for a single-generation exposure in a laboratory experiment, but Burggren's idea clearly envisages environmental exposure for multiple generations. Moreover, there are examples of epigenetic changes persisting for many generations in laboratory experiments. I will give just two examples here.

---

[40] W. Burggren, "Epigenetic Inheritance and Its Role in Evolutionary Biology: Re-Evaluation and New Perspectives," *Biology*, 2016, **5**(2): 24; doi: 10.3390/biology5020024.

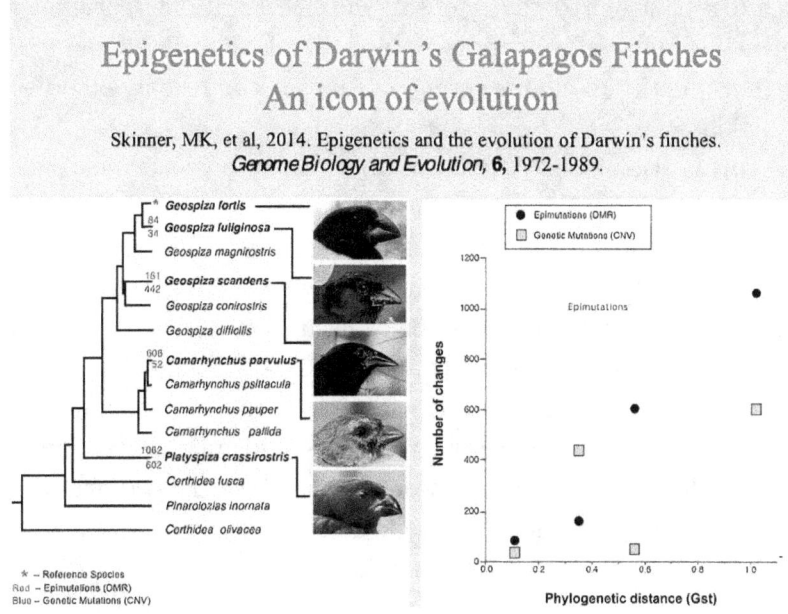

Figure 4.9: *Epigenetics and the Evolution of Darwin's Finches.*

The tiny planarian worm, *C. elegans*, is a favorite organism for genetic and molecular biological studies. It can be infected with a particular virus. Organisms that possess the correct DNA can react to this environmental stimulus by making an RNA that silences the virus, preventing it from using the host mechanisms for reproduction. By breeding these worms with others that do not have the relevant DNA, Oded Rechavi and his colleagues obtained worms in subsequent generations that do not have the relevant DNA. Yet, they still inherit the acquired resistance to the virus.[41] They do so by small quantities of the viral-silencing RNA passing through the male germ line to be amplified in each generation by an enzyme called RNA polymerase. The acquired characteristic is transmitted in this way through at least 100 generations. This example shows that the

---

[41] O. Rechavi, G. Minevich, and O. Hobert, "Transgenerational Inheritance of an Acquired Small RNA-Based Antiviral Response in *C. elegans*," *Cell*, 2011, **147**: 1248–1256.

idea that an acquired characteristic will necessarily die out after a few generations is not correct. It also reveals that RNAs can also be transmitted through the germ line. DNA is not the only inherited material.

This is also a good example of how modern epigenetic research is confirming Darwin's idea of "gemmules." RNAs can clearly play a role very similar to his explanation for Lamarckian inheritance.

Robust inheritance of an acquired epigenetic characteristic has been demonstrated in mice by Joe Nadeau's group in Seattle. They worked on a family of proteins that can insert mutations in DNA and RNA to show inheritance of epigenetic marking. This shows that the genome is not completely wiped clean of the marks in the germ line. On the contrary, Nadeau's work shows that such inheritance can be just as robust as standard genetic inheritance and can persist for many generations.[42] Epigenetic marking of the chromosome proteins has also been shown to be inherited.[43] The transmission of epigenetic marking has recently been shown to play a role in the inheritance of obesity in humans,[44] while the transmission of RNAs in sperm mediate the transmission of obesity in mice.[45]

Now I come to the question of rarity. As Burggren notes[46]

---

[42] V.R. Nelson, J.D. Heaney, P.J. Tesar, N.O. Davidson, and J.H. Nadeau, "Transgenerational Epigenetic Effects of the Apobec1 Cytidine Deaminase Deficiency on Testicular Germ Cell Tumor Susceptibility and Embryonic Viability," *Proceedings of the National Academy of Sciences, USA*, 2012, **109**: E2766–E2773; doi: 10.1073/pnas.1207169109.

[43] J.R. McCarrey, "The Epigenome—A Family Affair," *Science*, 2015, **350**: 634–635; and K. Siklenka, et al., "Disruption of Histone Methylation in Developing Sperm Impairs Offspring Health Transgenerationally," *Science*, 2015, **350**(6261): aab2006; doi: 10.1126/science.aab2006.

[44] I. Donkin, et al., "Obesity and Bariatric Surgery Drive Epigenetic Variation of Spermatozoa in Humans," *Cell Metabolism*, 2016, **23**: 369–378. (http://dx.doi.org/10.1016/j.cmet.2015.11.004)

[45] Q. Chen, et al., "Sperm tsRNAs Contribute to Intergenerational Inheritance of an Acquired Metabolic Disorder," *Science*, 2016, **351**(6271): 397–400; doi: 10.1126/science.aad7977; and U. Sharma, et al., "Biogenesis and Function of tRNA Fragments During Sperm Maturation and Fertilization in Mammals," *Science*, 2016, **351**(6271): 391–396; doi: 10.1126/science.aad6780.

[46] W. Burggren, "Epigenetic Inheritance and Its Role in Evolutionary Biology: Re-Evaluation and New Perspectives," *Biology*, 2016, **5**(2): 24; doi: 10.3390/biology5020024.

(see also Tollefsbol[47]), the examples of transgenerational epigenetic inheritance are fairly few, and long-term examples of the kind I show here are even rarer. These are early days in such research, so we don't yet know how rare the phenomena may be. All I will do here is to note that speciation is also rare. The rarity of long-term transgenerational epigenetic effects would not exclude their contributing to rare speciation.

4.8. The Gene-Centric View Claims Genetic Change is Always Random with Respect to Function

The claim that genetic change is always random with respect to function. It is almost certain that it would be, since randomness at the molecular level is what you would expect even if functionality exists at other levels.

I have already described the extent and possible role of stochasticity in item 2, above. There are several lessons to be learned from the experiments I described there:

1. At the molecular level, stochasticity is inevitable in organisms, just as it is in purely physical, non-living systems.
2. The extent of stochasticity in gene expression is controlled by the organism in a way that is inheritable.
3. Such inheritance can be transgenerational.

To these points I now add another:

The idea that there might be one-way, fully determinate causation between the genome and the phenotype goes back as far as Max Delbrück and Erwin Schrödinger. They proposed that the genetic molecular structure would be that of a determinate, aperiodic crystal. If we regard a polymer as a linear crystal, this is a good description

---

[47] T. Tollefsbol, ed., *Transgenerational Epigenetics: Evidence and Debate*. Waltham, MA: Academic Press, 2014.

of the genome. I think, therefore, that they were influenced by crystallography, which gives a determinate "read-out" of the atomic structure of a crystal. This led Schrödinger to propose that, while physics may be the generation of order (e.g., in thermodynamics) from disorder (random movements of gas molecules), biology would turn out to be the generation of order (phenotype) from order (genotype). As we have seen, that cannot be true. There is no way in which the molecules in living systems can be immune from molecular stochasticity. In his 1944 Dublin lectures, *What is Life?*, Schrödinger was clearly struggling to find a way around this problem:[48]

> We seem to arrive at the ridiculous conclusion that the clue to understanding of life is that it is based on a pure mechanism, a "clock-work" in the sense of Planck's paper.[49] The conclusion is not ridiculous and is, in my opinion, not entirely wrong, but it has to be taken "with a very big grain of salt."

He then explains the "very big grain of salt" by showing that even clockwork is, "after all statistical.")[50] My reading of these last pages of Schrödinger's book is that he realizes that something is not quite right, but is struggling to identify what it might be. We would now say that the molecules involved (DNA) *are* subject to statistical variation (copying errors, chemical and radiation damage, etc.), which are then corrected by the protein machinery that enables DNA to be a highly reproducible molecule. This is a three-stage process that reduces the error rate from 1 in 104 to around 1 in 1010, which is an astonishing degree of accuracy. The order at the molecular scale is therefore actually created by the system as a whole. This requires energy, of course, which Schrödinger called "negative entropy." Perhaps, therefore, this is what Schrödinger was struggling towards, but we

---

[48] E. Schrödinger, *What is Life?* (with *Mind and Matter* and *Autobiographical Sketches*). Cambridge: Cambridge University Press, 2012; p. 82.

[49] Schrödinger is here referring to Max Planck's article, "Dynamische und statistische Gesetzmäßigkeit," *Zeitschrift für Elektrochemie und angewandte physikalische Chemie*, 1917, **23**: 63.

[50] E. Schrödinger, *What is Life?* (with *Mind and Matter* and *Autobiographical Sketches*). Cambridge: Cambridge University Press, 2012; p. 83.

can only see this more clearly in retrospect. He could not have known how much the genetic molecular material experiences stochasticity and is constrained to be highly reproducible by the organism itself.

This resolution of Schrödinger's problem is very important. As we have seen in discussing the interpretation of the Central Dogma (see item 6, above), a determinate, one-way coding does not guarantee that information determining the patterns of gene expression cannot be passed from the organism to its genome. In fact, this is happening all the time.

Organisms, therefore, can be seen to control stochasticity, and even to impose order, when it is necessary to do so.

In transmitting DNA sequences from one generation to another, this is necessary and organisms employ whole armies of networks to ensure nearly faultless transmission of sequences. But there is no need for organisms to ensure determinate behavior at the molecular level for any other transmission of information. In particular, functionality at the higher levels does not require the imposition of order at lower levels. Randomness is then what we would expect to see. But it would be wrong to assume that this means that goal-directed functionality cannot be consistent with non-functional randomness at a molecular level. Indeed, organisms not only live with such randomness, they actually control its extent and distribution.

4.9. The Gene-Centric View Claims Neo-Darwinism is Obvious and Necessarily True

The claim that neo-Darwinism is obvious and necessarily true (often advanced by Dawkins, as in the debate with Margulis). If that were the case, it would become a tautology and not open to experimental verification—and therefore not much good as a scientific theory.

I discussed the issue of parsimony in item 5, above. The claim that neo-Darwinism is necessary is usually not far behind the claim that it is parsimonious. When Weismann wrote,[51]

> ...once it is established that natural selection is the only principle
> which has to be considered, it necessarily follows that the facts can
> be correctly explained by natural selection[,]

He was close to claiming explicitly that this is a necessary statement. The second phrase necessarily follows if the assumption in the first phrase is correct. The problem with necessary statements is that they are no longer open to experimental verification, unless, of course, some error of logic has crept it—in which case the statement is either not necessary (and might be falsified) or meaningless (which is even worse).

I don't myself think that the main original thesis of neo-Darwinism is a necessary statement since it must be open to experimentation whether variation really is random with respect to function. As we have seen, the experimental proof of that is at least questionable since randomness at a molecular level does not entail randomness at a higher level.

The more serious way in which the theory gets close to being unfalsifiable is the continuous process of extension. Some of the claims for the theory being extensible seem to me to be based on forgetting what the theory originally stated. If we go back to the original definitions of neo-Darwinism, it is fairly easy to see how it could be falsified.

At the top of this list is the absolute exclusion of the inheritance of acquired characteristics as championed by Darwin's great French predecessor, Jean-Baptiste Lamarck (right). It is impossible to read Weismann, Mayr, Dawkins, and many others in any other way. Moreover, reintroducing Lamarckian inheritance would take us back to Darwin. Neo-Darwinism would not then be necessary to distinguish the theory from Darwinism.

---

[51] A. Weismann, "The All-Sufficiency of Natural Selection: A Reply to Herbert Spencer," *Contemporary Review*, 1893, **64**: 309–338; p. 337.

This is the central issue in an exchange published recently in the *Journal of Experimental Biology*. My reply was:[52]

> If, as the commentator seems to imply, we make Neo-Darwinism so flexible as an idea that it can accept even those findings that the originators intended to be excluded by the theory it is then incumbent on modern Neo-Darwinists to specify what would now falsify the theory. If nothing can do this then it is not a scientific theory.

I think, therefore, that the ball is in the other court. What would an extended version of neo-Darwinism accept as falsification? Following all the extensions that have now been proposed, it is not clear to me whether there is any agreed definition of what the theory allows and doesn't allow. Different neo-Darwinists see this question differently.

4.10. The Gene-Centric View Appeals to Authority

The claim was formulated by some of the greatest scientists of the twentieth century, so it must be right. *Nullius in Verba!*

The neo-Darwinist Modern Synthesis was developed by very major figures in twentieth-century biology. No one can doubt that. The scientists who developed the Modern Synthesis were brilliant and they were among the most influential scientists of the twentieth century. They formulated the best hypothesis they could that would combine the observations and insights of Darwin and Wallace on the role of natural selection with the discoveries of Mendel and the idea of the Weismann Barrier on genetics. As a hypothesis, it was very successful. The study of the genetics of populations was transformed and became a rigorously mathematical discipline.

It is perfectly possible to acknowledge these achievements of neo-Darwinism while dissenting from the view that the theory encompasses all the processes that can contribute to evolutionary

---

[52] D. Noble, "Central Tenets of neo-Darwinism Broken. Response to 'Neo-Darwinism is Just Fine,'" *Journal of Experimental Biology*, 2015, **218**: 2659; doi: 10.1242/jeb.125526.

change. The distinction and authority of the formulators of a theory are not what matter.

In the UK, the Royal Society, which is the equivalent of the National Academy of Sciences in the U.S., has a Latin motto, NULLIUS IN VERBA,[53] which can be roughly translated as "don't take anyone's word for it."[54]

It is evidence that counts. My view is that the evidence strongly suggests that the time has come for a rethink. That view is what I develop more fully in my most recent book, *Dance to the Tune of Life: Biological Relativity* (Cambridge University Press, 2017).

4.11. Epilogue: The Language of Neo-Darwinism

This item was not in the original 10 items at the end of the interview, but I think it is important to conclude this statement of my position with some comments on the language of neo-Darwinism since the problems the theory faces in accommodating many experimental findings have their origin in neo-Darwinist metaphors and other forms of representation, rather than in experimental biology itself.

These colorful metaphors have been responsible for, and express, the way in which twentieth-century biology has most frequently been interpreted and presented to the public. In addition, therefore, to the need to accommodate unanticipated experimental findings, we need to review the way in which we interpret and communicate experimental biology. The language of neo-Darwinism and twentieth-century biology reflects highly reductionist philosophical viewpoints, the concepts of which are not required by the scientific discoveries themselves. In fact, it can be shown that, in the case some of the central concepts of neo-Darwinism, such as "selfish genes" or "genetic program," no biological experiment could possibly distinguish even between completely opposite conceptual interpretations

---

[53] Visible on its official seal—Ed.

[54] Literally, "on the word of no one"—Ed.

of the same experimental findings.[55] The concepts, therefore, form a biased interpretive veneer that can hide those discoveries in a web of interpretation.

I refer to a "web of interpretation" since it is the whole conceptual scheme of neo-Darwinism that creates the difficulty. Each concept and metaphor reinforces the overall mind-set until it is almost impossible to stand outside it and to appreciate how beguiling it is.

Since neo-Darwinism has dominated biological science for over half a century, its viewpoint is now so embedded in the scientific literature, including standard school and university textbooks, that many biological scientists may themselves not recognize its conceptual nature, let alone question incoherencies or identify flaws. Many see it as merely a description of what experimental work has shown: the idea in a nutshell is that genes code for proteins that form organisms via a genetic program inherited from preceding generations and which defines and determines the organism and its future offspring.

What is wrong with that? The problem is that the conceptual scheme is neither required by, nor any longer productive for, the experimental science itself. Nor is it consistent with the principle of relativity applied to multi-scale biology. That is why many scientists in the physiological sciences have great difficulty reconciling their

---

[55] As an example, consider the following two paragraphs:

"Now they swarm in huge colonies, safe inside gigantic lumbering robots, sealed off from the outside world, communicating with it by tortuous indirect routes, manipulating it by remote control. They are in you and me; they created us body and mind; and their preservation is the ultimate rationale for our existence. (R. Dawkins, The Selfish Gene [Oxford UP, 1976]; p. 21.)

Now they are trapped in huge colonies, locked inside highly intelligent beings, molded by the outside world, communicating with it by complex processes, through which, blindly, as if by magic, function emerges. They are in you and me; we are the system that allows their code to be read; and their preservation is totally dependent on the joy we experience in reproducing ourselves. We are the ultimate rationale for their existence. (D. Noble, The Music of Life [Oxford UP, 2006]; p. 12.)"

Apart from the obviously true statement "they are in you and me," the statements are diametrically opposed. Yet no conceivable experiment could distinguish between them. See D. Noble, "Neo-Darwinism, the Modern Synthesis, and Selfish Genes: Are They of Use in Physiology?," Journal of Physiology, 2011, 589: 1007–1015.

science with neo-Darwinism[56] and see the need for a new conceptual framework.[57]

As Waddington saw very clearly in his book *The Strategy of the Genes* (see the quote from this book in item 3, above), the language of neo-Darwinism leads to questionable ideas on what it is to be human. So, I finish by repeating this quote from *The Selfish Gene*:[58]

> Let us try to teach generosity and altruism, because we are born selfish. Let us understand what our own selfish genes are up to, because we may then at least have the chance to upset their designs, something that no other species has ever aspired to.

The empirical discoveries of biological science do not show that we are born selfish. The idea that we are formed of "selfish" DNA, even if it were correct, could not justify attributions of selfish or cooperative behavior at the level of the organism. My selfish or cooperative behavior (like all humans, I exhibit both) depends on my genes (or, rather, the gene products: proteins and RNAs) cooperating in vast biological networks in interaction with the contextual logic of my environment. It is a category mistake to confuse attributions at a molecular level with those at the level of the whole organism.

Descartes had the same problem when he saw the need to argue that humans are not just mechanisms. Like Dawkins, he also was compelled to consider that humans are unique, that they have a capacity "that no other species has ever aspired to." I cannot see how an evolutionary biologist can possibly accept this view since we evolved from animals. The abilities to empathize and love are clearly seen in animals other than humans.

The mistake lies in regarding animals as pure mechanisms. That is what leads to special pleading by Dawkins, as much as by Descartes, to somehow regard humans as the only exceptions.

---

[56] D. Noble, "Neo-Darwinism, the Modern Synthesis, and Selfish Genes: Are They of Use in Physiology?," *Journal of Physiology,* 2011, **589**: 1007–1015.

[57] D. Noble, "Evolution Beyond neo-Darwinism: A New Conceptual Framework," *Journal of Experimental Biology,* 2015, **218**: 7–13.

[58] R. Dawkins, *The Selfish Gene.* Oxford: Oxford University Press, 1976; p. 3.

The Czech novelist Milan Kundera expressed the connection between a mechanistic view of animals and the consequent difficulties for our concept of humanity when he wrote this passage in his brilliant book, *The Unbearable Lightness of Being*:[59]

> That is why it is so dangerous to turn an animal into a *machina animata*, a cow into an automaton for the production of milk. By so doing, man cuts the thread binding him to paradise and has nothing left to hold or comfort him on his flight through the emptiness of time.

4.12. Acknowledgment

This text reflects some of my articles published over the last 10 years (see The Music of Life Sourcebook[60]) and my book, *Dance to the Tune of Life: Biological Relativity*. Where sections of text have been borrowed from those publications, they have been extensively rewritten for this Dialogue.

---

[59] M. Kundera, *The Unbearable Lightness of Being*. New York: Harper & Row, 1984; p. 297. (Written in Czech in 1982, but originally published in French in 1984.—Ed.)

[60] [http://www.musicoflife.website/pdfs/The%20Music%20of%20Life-sourcebook.pdf.]

5. David Sloan Wilson's Response: Tinbergen's Four Questions Concisely Describe a Fully Rounded Evolutionary Perspective

While reading Denis Noble's contribution to this Dialogue and reviewing my own, I found myself wondering what a reader encountering the ideas for the first time might be thinking. I worry that such a reader might feel bewildered and ultimately bored by who said what and whether it was well received or ignored. Can't we just move ahead with the best of our current knowledge?

More interesting is the question of why science moves rapidly along some fronts and slowly along others. Everyone can agree that the pace of scientific change is highly variable. If we can diagnose the causes of variability in the past, then perhaps we can accelerate the pace of scientific change in the future. It is in this spirit that the history of evolutionary thought becomes interesting and relevant to those encountering the subject for the first time.

It is important to keep in mind that Professor Noble and I are defining neo-Darwinism in different ways. My definition is based on a four-fold distinction made by Niko Tinbergen in 1963, which states that a fully rounded evolutionary perspective requires paying attention to function, mechanism, development, and phylogeny.[1] "Tinbergen's four questions" can be mapped onto a two-fold distinction between ultimate and proximate causation made independently by Ernst Mayr in 1961.[2]

Roughly, Tinbergen's "function" and "phylogeny" questions map on to Mayr's "ultimate causation," and Tinbergen's "mechanism" and "development" questions map onto Mayr's "proximate causation." The claim I defend is that Tinbergen's four questions remain a concise description of a fully rounded evolutionary perspective and that to suggest otherwise is a distraction.

Noble defines neo-Darwinism in terms of a number of positions that emerged during the twentieth century: that genes are the only mechanism of inheritance; that genetic variation is random with respect to what is selected; that germ cells are isolated from somatic cells; and so on. Positions such as these have a way of hardening into dogmas that resist change for decades, and to some extent still today. I do not contest these facts and I am a veteran of another position that hardened into a dogma: that natural selection always operates at the level of the individual and/or the gene.

It is fine for us to define neo-Darwinism in different ways. As long as we are clear about our differences, then we are like two mountaineers viewing the same mountain from different angles to plan an ascent. I like my angle because it has great current utility. If Tinbergen's four questions remain good enough, then they can be easily taught and kept in mind by beginner and expert alike. Against this background, I will review Professor Noble's 10 reasons why from his perspective "neo-Darwinism is not enough," with three questions in mind. First, are his 10 reasons legitimate? Second, do they challenge my claim that Tinbergen's questions are good enough? Third, why didn't scientific inquiry proceed more quickly and smoothly for these topics?

---

[1] N. Tinbergen, "On Aims and Methods of Ethology," *Zeitschrift für Tierpsychologie*, 1963, **20**: 410–433.

[2] E. Mayr, "Cause and Effect in Biology," *Science*, 1961, **134**(3489): 1501–1506.

1. Major diseases still plague humanity. Professor Noble correctly notes that the Human Genome Project largely failed to deliver on its grandiose claims. It turns out that knowing the DNA sequence of organisms does not provide sufficient information to explain more than a small fraction of variation for phenotypic traits such as body height or susceptibility to a disease. This is because most phenotypic traits are highly polygenic and influenced by gene-gene and gene-environment interactions operating though complex developmental and physiological systems.

Does this pose a challenge to Tinbergen's four questions? Not in the least, since they call upon us to study mechanisms and development. Then why were the architects of the Human Genome Project so naïve as to think that phenotypic traits map directly onto genes? In part because they were selling something to Congress. In part because of the appeal of reductionism, which manifests itself for many topics in science, not just evolution. I have noticed for the study of ecology, for example, that holistic descriptions of the tapestry of nature in the 1950s gave way to simple models of single species dynamics and multispecies interactions. These models were somewhat explanatory, but also revealed their own weaknesses. Complications were added and now nature is described as a grand tapestry, after all. Of course, the current description is mechanistically far richer than the earlier description.

Is it necessary for the study of proximate mechanisms to pass through a simplistic phase before embracing more complex explanations that were always known to have existed? I don't know, but I agree with Professor Noble that the study of physiological systems was neglected and deserves the kind of support that the Human Genome Project received.

2. Privileging any one level in biological systems cannot be justified. I worry that Professor Noble is unduly influenced by a few provocateurs such as James Watson and Richard Dawkins. Watson's statement "there are only molecules—everything else is sociology" is tantamount to saying, first, that the mechanism question is the only one worth asking, and, second, that the mechanism question can be answered entirely at the molecular level. I agree with Professor Noble that both of these claims are stupid, but I also don't think that Watson's quip represents neo-Darwinism or probably even Watson in more sober moods.

There are two major forms of holism that are already part of the neo-Darwinian synthesis, even as formulated by Professor Noble. The first is ultimate causation (=Tinbergen's function and history questions). The holistic statement "the parts permit but do not cause the properties of the whole" is literally correct when environmental forces operate on heritable variation.[3] That is why the distinction between proximate and ultimate causation is so insightful.

The second form of holism is purely mechanistic and establishes that higher-level physical entities, such as molecules, have properties that cannot be reduced to the properties of their parts, such as their constituent atoms. This point can be made for compounds such as salt and water without requiring the biological examples that Professor Noble provides. It is so firmly established that one article (Sober 1999) begins, "If there is now a received view among philosophers of mind and philosophers of biology about reductionism, it is that reductionism is mistaken."[4]

An important lesson to learn is that a scientist such as James Watson can be brilliant and well-read in some respects and painfully ignorant in other respects. A distinguished scientific reputation is not (or should not be) a license to hold forth on all subjects.

To conclude, Professor Noble's second critique is valid, but doesn't fairly represent the neo-Darwinian synthesis, even as he

---

[3] D.S. Wilson, "Holism and Reductionism in Evolutionary Biology," *Oikos*, 1988, **53**: 269–273.

[4] E. Sober, "The Multiple Realizability Argument Against Reductionism," *Philosophy of Science*, 1999, **66**: 542–564; p. 542.

defines it, and affirms rather than challenges the four-question perspective.

3. The gene-centric view has damaging consequences. Alas, I'm
afraid I must agree with Professor Noble that he is not a social
historian and is making statements on a par with Watson's quip about
molecules. To begin, some of the disturbing implications of evolution
began with Darwin, not neo-Darwinism. The Christian worldview
imagines the universe as harmonious from top to bottom, from the
smallest insect to the stars in heaven. Darwin's theory explained
functional design at the level of individual organisms, but required
special conditions (group selection) to explain functional design
even at the level of single-species social groups, much less higher
levels of nature such as ecosystems and the biosphere. Gene-centrism
had nothing to do with these and other disturbing implications of
Darwin's theory for the Christian worldview.

While Darwinism was genuinely threatening to some worldviews, the idea that it unleashed a plague of toxic policies justifying
the strong taking from the weak is largely false. I speak with authority
on the subject, having organized a special edition of "This View of
Life" titled "Truth and Reconciliation for Social Darwinism." The
conjecture that evolutionary theory's turn toward gene-centrism had
additional pernicious effects is even more far-fetched. I encourage
Professor Noble to read the special edition before writing anything
more on this subject!

The idea that we are born selfish and must be taught altruism
goes back millennia and has been expressed in many different forms,
including the Christian concept of original sin. It's true that Dawkins
and others such as G.C. Williams,[5] Richard Alexander,[6] and Michael

---

[5] G.C. Williams, *Adaptation and Natural Selection: A Critique of Some Current Evolutionary Thought*. Princeton: Princeton University Press, 1966.

[6] R.D. Alexander, *Darwinism and Human Affairs*. Seattle: University of Washington
Press, 1979.

Ghiselin[7] gave the idea an evolutionary spin in the 1960s and 1970s. It's probably not a coincidence that this development in evolutionary biology coincided with the rise of methodological individualism in the social sciences, rational choice theory in economics, and the rise of individualism in popular culture. The trend in evolutionary biology was more likely a manifestation of these other trends than the cause. Nor was gene-centrism required to create an individualistic worldview. Friedrich Hayek,[8] one of the main architects of neoliberal economic thought, was an early adopter of cultural group selection theory, which did not prevent him from becoming individualistic in his own way.

These topics have little to do with neo-Darwinism, even as defined by Professor Noble, and they don't threaten the four-question perspective, but they do shed some light on why scientific controversies can be protracted—because they are thoroughly entwined with non-scientific, cultural worldviews. Please see my own essay for more on multilevel selection as a protracted controversy.

4. The gene-centric view resists new findings. I am puzzled by the first paragraph of this section. Far from resisting a new finding, a supporter of neo-Darwinism accepted the inheritance of acquired characteristics, but for Professor Noble this was a contradiction in terms.

The reason that directed forms of evolution can be reconciled with neo-Darwinism is because they can be shown to evolve from undirected forms of evolution. Once this point is understood and mechanisms are identified, then there is no barrier to accepting directed forms of evolution from a neo-Darwinian perspective. Individual learning and cultural evolution provide clear examples of directed evolutionary processes. If genetic and epigenetic processes are also discovered, then they will be of the same type.

---

[7] M.T. Ghiselin, *The Economy of Nature and the Evolution of Sex.* Berkeley: University of California Press, 1974.

[8] F.A. Hayek, *The Fatal Conceit.* London: Routledge, 1988.

That said, I agree with Professor Noble that many evolutionary biologists are dogmatically opposed to anything that smells like Lamarckism, just as others are dogmatically opposed to anything that smells like group selection. Why should this be? In part because they are often taught about these subjects in a dogmatic fashion. Much as we might want science education to be open-minded, it often takes the form of identifying what is taboo and what is permitted, even at the college and post-graduate levels. This is a problem for science education in general, which is by no means restricted to the gene-centric view.

5. The gene-centric review claims parsimony. I am familiar with this claim because parsimony played a large role in the rejection of group selection.[9] The idea was that individual-level selection is somehow simpler than group-level selection and therefore preferable in the absence of other deciding information. I agree that this is an extremely weak argument for most topics in science, as Elliott Sober has recently recounted in detail.[10] It is especially weak within evolutionary theory, where adaptations are expected to be more like what a tinkerer, rather than an engineer, would build.

While I side with Professor Noble on the topic of parsimony, I still must object to this passage: "Whole areas of economics, sociology, and philosophy are based on interpretations of selfish gene viewpoints." I think that this passage greatly overstates the influence of gene-centric thinking in these other disciplines. See my interview with the sociologist Russell Schutt titled "Why Did Sociology Declare Independence from Biology (and Can They Be Reunited)?"[11]

---

[9] G.C. Williams, *op. cit.*; for a critique, see E. Sober and D.S. Wilson, *Unto Others: The Evolution and Psychology of Unselfish Behavior.* Cambridge, MA: Harvard University Press, 1998.

[10] E. Sober, *Ockham's Razors: A User's Manual.* Cambridge: Cambridge University Press, 2015.

[11] [https://web.archive.org/web/20170502040118/https://evolution-institute.org/articl e/why-did-sociology-declare-independence-from-biology-and-can-they-be-reunited-an-int erview-with-russell-schutt/.]

(which is part of the Truth and Reconciliation Special Edition), for a more informed view of that discipline in relation to evolutionary theory.

6. The gene-centric view claims to settle the question of Lamarckism. In this and other sections, Professor Noble reviews the work of figures such as James Mark Baldwin, C.H. Waddington, Barbara McClintock, and Lynn Margulis, to which we could add figures such as Sewall Wright and concepts such as niche construction and developmental systems theory. All of these people and concepts were influential enough to be remembered—what student of evolution hasn't seen an image of Waddington's developmental landscape or Wright's adaptive landscape?—but somehow failed to occupy center stage.

Why should this be?

In part, because they embraced the notion of complex interactions, which are inherently more difficult to study than more simple interactions. I have some sympathy with parsimony as a research strategy, which attempts to explain as much as possible with simple models before resorting to more complex models. It is instructive to compare the three main pioneers of theoretical population genetics: Ronald Fisher; J.B.S. Haldane; and Sewall Wright. All three were gene-centric, the difference being that Wright was more drawn to study complex genetic interactions that result in multiple local equilibria (his shifting balance theory), whereas Fisher and Haldane began by modeling genes with additive effects. Fisher and Haldane were arguably able to make more rapid progress by picking the "low-hanging fruit" with their simple models. Hence, they had a greater influence at first, but once the low-hanging fruit had been picked, then Wright's work became increasingly important. If all of this takes place in the span of a few decades, then it is hard to fault the scientific process for being too biased or too slow.

Another point is that if it's possible to make an end run around complexity, then by all means one should do so. Adaptationist thinking (Tinbergen's function question) is one example of an end run. By assuming heritable phenotypic variation, it makes progress without needing to understand the underlying mechanisms. As another example, imagine comparing the phenotypes of two genotypes that differ by only a single mutation. Each allele influences the phenotype through a complex web of developmental and physiological interactions, but these interactions can be ignored if we merely want to select the mutant phenotype. If rapid progress along these lines took place in the mid-twentieth century, with topics such as evo-devo only coming into their own during late in the century, should we be too surprised?

Another point is that a few decades is not a long time. Science is a process of constructive disagreement and several rounds of hypothesis testing might be required to dislodge someone's firmly held view. Professor Noble states that "the 1956 version of the Central Dogma had to be substantially revised in the 1970's when it was discovered that the step from DNA to RNA is in fact reversible." Two decades is an extraordinarily short time for this kind of scientific progress!

7. The gene-centric view claims that epigenetic inheritance is short-lived. Scientists can be conservative or progressive, much like politicians or religious believers. Some will rush to say "This changes *everything!*," while others will never change their mind, especially if they have received a dogmatic education. This is true for all topics in science and is not restricted to evolutionary theory.

If we think of epigenetic inheritance as a form of multigenerational phenotypic plasticity, then it is helpful to review what we know about intragenerational phenotypic plasticity.

The function of phenotypic plasticity is to track environmental changes during the lifetime of the organism. The exact form of phenotypic plasticity, therefore, depends upon the pattern of environmental changes. Arctic mammals change their coat color with the season. A turtle withdraws into its shell when a predator approaches and comes back out when the predator leaves. Some forms of phenotypic plasticity are closed (involving a fixed repertoire of behaviors), while others are open (involving a variation and selection process such as operant conditioning).

For the purpose of argument, let's assume that there are no mechanistic barriers preventing the phenotypic state of the parent to be transmitted to its offspring. If so, then we'd expect the same diversity of patterns for transgenerational phenotypic plasticity that we see for intragenerational phenotypic plasticity. The duration of phenotypic change would be calibrated to the frequency of environmental change for each particular trait.

According to Jablonka and Lamb,[12] many forms of transgenerational phenotypic plasticity mediated by epigenetic effects are relatively closed, such as the fast and slow life history strategies that are expressed in rats depending upon how much they are licked by their mothers.[13] These hold little potential for cumulative evolutionary change. However, behaviors that are learned by operant conditioning during one generation and transmitted across generations by various forms of social learning are more open-ended and cumulative. Forms of symbolic thought that are distinctively human have evolved into a full-blown inheritance system that operates in parallel to the genetic inheritance system.[14]

As I have emphasized elsewhere in this Dialogue, some of the most radical claims that are being advanced for genetic and epigenetic mechanisms are similar in kind to claims that have already been established for individual learning and transgenerational cultural evolution. Going back to basics and centering evolution on the concept of heredity, not merely genes, is revolutionary against the

[12] E. Jablonka and M. Lamb, *Evolution in Four Dimensions: Genetic, Epigenetic, Behavioral, and Symbolic Variation in the History of LIfe*. Cambridge, MA: MIT Press, 2006.

[13] N.M. Cameron, D. Shahrokh, A. Del Corpo, S.K. Dhir, M. Szyf, F.A. Champagne, and M.J. Meaney, "Epigenetic programming of phenotypic variations in reproductive strategies in the rat through maternal care," *Journal of Neuroendocrinology*, 2008, **20**(6): 795–801.

[14] R.A. Paul, *Mixed Messages: Cultural and Genetic Inheritance in the Constitution of Human Society*. Chicago: University of Chicago Press, 2015.

background of gene-centric evolutionary theory, a point on which Professor Noble and I agree. Tinbergen's four questions are needed for everything that counts as an evolutionary process, no matter what the particular mechanism of inheritance.

8. The gene-centric view claims genetic change is always random with respect to function. Much of what Professor Noble writes in this section affirms this view, adding that randomness is also controlled and incorporated into function. Behavioral examples can be added to the physiological examples. Almost all animal and human decision-making involves a component that is random with respect to what is selected. The same is true for evolutionary algorithms in computer science. Where else can novelty come from? None of this is a radical departure from neo-Darwinism, even as defined by Professor Noble, nor does it challenge the four questions perspective.

9. The gene-centric view claims neo-Darwinism is obviously and necessarily true. Neo-Darwinism, as defined by Professor Noble, is a collection of empirical claims. Each claim is acknowledged by all to be empirically falsifiable, the quotes by Weissman and Dawkins notwithstanding. When Weissman chopped off the tails of mice, he was prepared to accept the results if the next generation of mice had shorter tails.

A neo-Darwinist can accept Lamarckian inheritance while remaining a neo-Darwinist if the mechanism of Lamarckian inheritance can be shown to evolve by a neo-Darwinian process. Since that is the case for all known Lamarckian inheritance mechanisms, then Lamarckian inheritance has been reconciled with neo-Darwinism. Some neo-Darwinists can be faulted for failing to acknowledge Lamarckian inheritance or for trying to marginalize its importance, but it is not helpful for Professor Noble to declare that Lamarckian inheritance lies outside the orbit of neo-Darwinism by definition.

10. The gene-centric view appeals to authority. I began my own essay by saying that I was an odd choice to defend the position "Is neo-Darwinism enough?," since I am a well-known opponent of selfish gene orthodoxy and a booster of the ideas that sail under the name "Extended Evolutionary Synthesis." Hence, I admire Professor Noble's systems approach to physiology and share his holistic worldview.

Nevertheless, when I read his essay in detail in preparation for writing this response, I became increasingly unhappy with his critique of neo-Darwinism as he chooses to define it. His parting shot is especially cheap. Professor Noble seems to have worked himself up into such a "us vs. them" lather that he can't even grant that his opponents also function as scientists and, as a group, don't appeal to authority any more or less than other scientists.

5.1. Conclusion

In my own essay, I chose to defend the position that Tinbergen's four questions provide a concise description of a fully rounded evolutionary perspective that remains good enough for student and expert alike. To claim otherwise is a distraction, especially since broad swaths of scientific inquiry do not employ a fully rounded approach, even in the biological sciences.

In this response to Professor Noble's essay, I attempted to answer three questions. First, how successful is his critique of neo-Darwinism, as he chooses to define it? Second, does his critique challenge neo-Darwinism as I define it (Tinbergen's four questions)? Third, what can we learn from the history of neo-Darwinism about the factors that accelerate or impeded scientific progress?

My conclusion for the first question is mixed. I am broadly sympathetic with Professor Noble's systemic and holistic view of evolution, but many aspects of his critique of neo-Darwinism, as he defines it, are highly problematic.

My conclusion for the second question is that Tinbergen's four questions emerge unscathed. They are not challenged, and are frequently affirmed, by Professor Noble's 10 points.

My conclusion for the third question is that the history of neo-Darwinism, as defined by Professor Noble, includes many examples of science progressing more slowly than we might like. Here is a list:

- Selling science to funders requires false and boastful claims
- The appeal of reductionism
- Scientific worldviews are bound up with cultural worldviews
- Science education that is more dogmatic than we might like
- Appreciation of complex interactions comes after discovering the limitations of explanations based on simple interactions
- A few decades is not a long time for science as a process of constructive disagreement

What's striking about the list is that none of the items has much to do with neo-Darwinism per se. All are likely to impede progress for any scientific subject.

Although I have been critical of Professor Noble in some respects, I agree with him that evolutionary science will make major advances during the twenty-first century, especially for the study of evolution in relation to human affairs. As I see it, this is because the fully rounded, four-question approach will become much more widespread than it is now.

6. Denis Noble's Response: A Framework is Not a Testable Theory

My initial reaction to being paired with David Sloan Wilson in this dialogue was to wonder what we could possibly disagree about. Prof. Wilson has been at the forefront of revealing the limitations and conceptual difficulties of "selfish gene" theory, which many people (including Dawkins himself) actually regard as synonymous with neo-Darwinism.[1] Wilson's book, with the philosopher Elliot Sober, *Unto Others* (Harvard UP, 1998), is a landmark publication, widely acknowledged as such both in biology and in economics. The smaller sequel, *Does Altruism Exist?* (Yale UP, 2015), is also an important book. I did not know of *Unto Others* when I wrote *The Music of Life* (Oxford UP, 2006), but I wish I had.

Incidentally, like Wilson, I also believe that research in economics and management theory needs to take account of the new trends in biology. That is why I am one of the organizers of a three-day meeting of the Royal Society and the British Academy on "New Trends in Evolutionary Biology,"[2] which involves leading researchers from the sciences, humanities, and social sciences, and which will be published in the Royal Society journal, Interface Focus. I also

---

[1] It is worth adding that in the last 10 years I have noticed that many evolutionary biologists no longer think that the "selfish gene" represents their case.

[2] [https://royalsociety.org/science-events-and-lectures/2016/11/evolutionary-biology/.
]

lead a relevant project of the Balliol Interdisciplinary Institute, "A Systems Approach to Management Studies," which is investigating the ways in which the systems approach in biology can inform such approaches in economics and management.

6.1. Possible Relevance to Today's Globalized Economy

We are therefore largely on the same wavelength on many of the crucial issues facing both the biological and the social sciences today. These issues are urgent, and pressingly so. It may be far too simplistic to link selfish gene theory to the development of naively reduction-ist forms of neo-liberalism, and I know that my Oxford colleague Richard Dawkins would certainly not sanction such applications of his theory. But, for good or ill, others have done so. One might therefore go so far as to say that the biological reality or otherwise of altruism is one of the crucial issues of the twenty-first century.

That is highlighted by the fact that the world is coping with the consequences of catastrophic breakdown in our financial systems, with rapidly growing disparity between rich and poor, and clear evi-dence of dubiously ethical, even if justified as strictly legal, behavior by far too many people in the top echelons of power in commerce, industry, politics, sport… you name it. The revelations have become so frequent in the media and on the Internet that the world has almost become weary of, and no longer surprised by, the scandalous nature of what emerges. Again, I emphasize that it is a far cry from these revelations to drawing any firm conclusions about the influence of theory in science and economics on people's behavior. I agree with Professor Wilson's caution in this regard. But, at least the possibility is there staring us in the face. I referred in my earlier contributions in this dialogue to the influence of some gene-centric interpretations of biology on the eugenics movement in the twentieth century, leading, as we all know, to the horrors of the Second World War, with its undercurrent of the idea of "purifying" the human race. The eugenics movement had precisely this as one of its aims.

In his Statement, Wilson argues very forcefully that

> Darwin's theory did not unleash a plague of toxic social policies justifying inequality. At most, it was added to a quiver that was already full of other arrows, including religious and economic justifications. The claim that Darwin's theory was used to justify Hitler's war policies, either directly or indirectly, is demonstrably false...

But the relevant possibility is surely not so much that Darwin's theories were the key driver, but rather that the later eugenics movement, spearheaded by Francis Galton, Karl Pearson, and Walter Weldon, did provide clear biological arguments for "purifying" the human race. It is significant that it was only after the Second World War that my *alma mater*, University College London, dropped the title Professor of National Eugenics—a chair that had been founded and personally financed by Francis Galton. Galton's chair then became simply the Professorship of Human Genetics.[3]

Moreover, such dangerous interpretations of genetic science have been sufficiently powerful that proponents of the selfish gene view of neo-Darwinism are themselves aware of it and try to counter and distance themselves from such possibilities. As I noted in my Major Statement, Dawkins himself had to write as long ago as 1976, "Let us try to teach generosity and altruism, because we are born selfish." It is hard to read this statement as anything less than a fear that selfish gene theory might encourage selfish behavior by providing its biological justification.

But what if the assumption that we are "born selfish" simply isn't justified from a biological viewpoint? That is the conclusion I came to in an analysis of selfish gene theory that I published in 2011.[4] The essence of the article is that:

---

[3] [https://en.wikipedia.org/wiki/Galton_Laboratory.]

[4] D. Noble, "Neo-Darwinism, the Modern Synthesis and Selfish Genes: Are They of Use in Physiology?," *Journal of Physiology*, 2011, **589**: 1007–1015.

1. Selfish gene theory is confused about the definition of a gene in ways that are critical to the distinction between replicators and vehicles that selfish gene theory uses,

2. Attributing selfishness to a genome sequence is not a testable theory, and

3. In any case, selfishness is a level-dependent concept. Selfishness at one level does not guarantee selfishness at other levels.

That is why the debate about altruism in general—and this Dialogue in particular—are so important.

I was intrigued, therefore, to see how Professor Wilson would rise to what I thought would be the almost impossible task of defending the self-sufficiency of neo-Darwinism, at least as I and many others understand the term. I have three reasons for being so intrigued.

First, because orthodox neo-Darwinists have systematically refused to engage formally in the discussion. It is over 15 years since my book, *The Music of Life,* was published by Oxford University Press. The challenge to neo-Darwinism was already very clear in that book. That is precisely why so many reviewers of it have been so positive about its message.[5] Yet, not a single orthodox neo-Darwinist has answered it.[6] The same silence has greeted the substantial series of peer-reviewed articles I have published since and which form the basis of The Music of Life Sourcebook. This sourcebook, in turn, has formed the basis of *Dance to the Tune of Life,* my new book on what I am calling "biological relativity." That book greatly extends the source material.

My second reason for being intrigued is that self-sufficiency is precisely how the founder of neo-Darwinism, August Weismann, characterized his theory. In a published debate with Herbert Spencer in 1893 he claimed not only that neo-Darwinism was all-sufficient ("*Die Allmacht der Naturzüchtung*"), but also that it was necessar-

---

[5] There are 25 reviews on Amazon.com, with an average score of 4.8. None of them defends neo-Darwinism against the thesis of the book.

[6] Other than through insulting and inaccurate comments on blogsites.

ily so.[7] I now realize through this Dialogue that Weismann's and Wilson's uses of the concept of "self-sufficiency" cannot be the same. I will return to this question later because it is the central divide between me and Wilson.

To anticipate the issue briefly, I think that what Wilson and I see as the central features characterizing neo-Darwinism are *fundamentally* different. Wilson characterizes neo-Darwinism in terms of a necessary research *framework*—which is based on Nikolaas ("Niko") Tinbergen's (1963) four questions.[8] By contrast, I characterize neo-Darwinism as the specific and testable hypothesis originally formulated by August Weismann and Alfred Russel Wallace. The first, the research framework, is not falsifiable. It is rather an appeal for a correct methodology in ethology and related fields of science, and I largely agree with the methodology. I say that it is not falsifiable because there is no experiment that could falsify it. It could only be "falsified," in a rather different sense of that word, if the experience of many researchers using the proposed four question framework showed that it doesn't or cannot work. It would, I believe, take generations of scientists to reach that conclusion, and I even doubt whether that could ever be established. We can't know whether a general framework of methodology works or not without trying it, and in science that can take a very long time. It took 30 years, for example, to realize that sequencing genomes would not reveal "The

---

[7] The title of Weismann's response to Herbert Spencer, *Die Allmacht der Naturzüchtung* (Jena: Fischer, 1893), means "the all-sufficiency of natural selection." He wrote: "We accept it [*Allmacht*]…simply because we must, because it is the only plausible explanation that we can conceive." He admitted that it was not possible to observe the process in detail, so there could be no experimental proof, but continued "It does not matter whether I am able to do so or not, or whether I could do it well or ill; once it is established that natural selection is the only principle which has to be considered, it necessarily follows that the facts can be correctly explained by natural selection." For further details on the history of these nineteenth-century debates, see Stephen Jay Gould, *The Structure of Evolutionary Theory* (Harvard UP, 2002); pp 170–250.

[8] N. Tinbergen, "On Aims and Methods of Ethology," *Zeitschrift für Tierpsychologie*, 1963, **20**: 410–433.

Book of Life."[9]

By contrast, Weismann's original theory forming the foundation of neo-Darwinism is not only falsifiable, it has actually been falsified. As I will show in my conclusions, I believe that Professor Wilson and I are actually largely in agreement on these points, though he uses different ways of expressing them.

6.2. Tinbergen's Four Questions and Wilson's Framework for Research

I first encountered Tinbergen's work 50 years ago. In 1966 I was a young academic at Oxford University and had recently won my faculty appointment there on the basis of two 1960 publications in *Nature*, the first being an experimental discovery of two very different potassium channels in the heart, and the second being a mathematical model to show how this discovery helped to explain electrical activity in the heart and the existence of cardiac rhythm.[10] It was thought, therefore, by the Oxford University science faculty that I was a great rarity in those days: a biologist who could cope with advanced mathematics. Little did they know that, six years earlier, when I struggled to learn enough maths and computer programming to use the vast, but extremely slow, machines called "computers" in those days, I was far from competent in advanced mathematics, or even in quite ordinary mathematics.

I was greatly intrigued by a challenge put to me by the Faculty at Oxford to examine a doctoral thesis of a young research student who was using Laplace transforms to analyse sequence data in ethology experiments done under the supervision of Niko Tinbergen. I didn't know then, of course, that Tinbergen would, a few years later, win a Nobel Prize in Physiology or Medicine, together with Konrad Lorenz

---

[9] See Michael Joyner's lecture, "Chasing Mendel: Thinking About Genotype vs. Phenotype" (University of Oxford, 2016). [https://www.youtube.com/watch?v=QsOeRTa4fYs .]

[10] This part of my career is described in more detail in the Interview.

and Karl von Frisch. But I did know of Tinbergen's great reputation in following Lorenz to bring great rigor to the study of bird behavior. The idea, essentially, was that behavior is an evolvable trait just as much as an organ of the body is an evolvable entity. This was an important insight. The evolution of behavior can be studied scientifically from an evolutionary perspective just as much as the evolution of organs, or any other structure.

Behavior is a process, not a structure, but as I have argued in *The Music of Life*, and again in *Dance to the Tune of Life*, even non-behavioral processes in cells, organs, and tissues are what actually evolve generally. Heart rhythm, for example, is a process, not a structure, even though it depends on the microscopic and macroscopic structures of the heart and the organs and systems with which it interacts. The big problem with gene-centric views of biology was to mistake the structure (the DNA sequence) for the process. And that mistake gets hidden by reference to what are called "gene" circuits and "genetic" programs. Those circuits and programs necessarily involve much more than DNA sequences. They could just as readily be characterized as "cell," "organ," or "system" circuits or programs. Except that, when we take that into account, the program becomes the functionality. The need for reference to a separate "program" disappears.

This digression explains why I was intrigued by the invitation of the Oxford Faculty in 1966. Here was a doctoral thesis project in ethology supervised by one of the acknowledged founders of the field, in which the student was using mathematics to analyze his results. So, I bought a hefty and expensive tome to bone up on Laplace transforms. I have never regretted that decision. It led 10 years later to the publication of one of my most successful and most mathematical books, still cited frequently today, 40 years on.[11] But I confess that, 50 years later, I don't remember much of the thesis, other than that it

---

[11] J.J.B. Jack, D. Noble, and R.W. Tsien, *Electric Current Flow in Excitable Cells.* Oxford: Oxford University Press, 1976. This book has been cited nearly 2000 times and is still very influential as the standard text on the mathematics of electrophysiology.

fully deserved the doctorate for which it was submitted. But I can't forget who the student was, for he became one of the most successful science writers of all time. He was Richard Dawkins—yes, the same person already referred to in this Dialogue!—whose writing skills are legendary. I don't go along with "selfish gene" theory, as I know he knows well, but I greatly admire the skill in communicating his ideas and enthusiasm to the general public. Countless schoolchildren must have been drawn into science through his books. He and I interacted well and courteously in a now legendary debate between him and Lynn Margulis in Oxford in 2009, which I was privileged to chair.

Wilson's use of Tinbergen's four questions therefore rings many bells with me. So also does reference to Konrad Lorenz. For it was at the Konrad Lorenz Institute in Vienna in 2013 that I first came across the "extended evolutionary synthesis" (EES), when I was invited by Gerd Müller to lecture there. My lecture used the title "The Music of Life and the billion-year dance of the genes," so the concepts that developed to become *Dance to the Tune of Life* were already formulated three years ago. Later, when writing one of my recent articles on evolutionary biology, I used the book which emerged from the EES meeting[12] to characterize what I saw as the main differences between the EES and what I have called an "integrative, relativistic" view of evolutionary biology.[13]

6.3. Is the "Extended Evolutionary Synthesis" Sufficient?

I see the EES as having identified a set of processes in evolution that were not included in either Darwinism or neo-Darwinism. That was a very valuable advance and I have used the book that emerged from the EES meeting many times since. But I have also wondered why the

---

[12] M. Pigliucci, and G.B. Müller, eds., *Evolution, The Extended Synthesis.* Cambridge, Mass: MIT Press, 2010.

[13] D. Noble, "Evolution Beyond neo-Darwinism: A New Conceptual Framework," *Journal of Experimental Biology,* 2015, **218**: 7–13.

word "extended" was chosen. I am very sympathetic to extensions of theory in science. But I also think that a fundamental rethink is sometimes indicated. I believe that evolutionary biology has reached that stage. The diagrams reproduced below explain the difference between the EES, and my view that, while appreciating the motives and achievements of the EES, we should go beyond it.

The problem with neo-Darwinism is that it became hardened into a strongly dogmatic insistence that certain processes that prede-cessors (including notably both Lamarck and Darwin) had proposed to include in evolutionary mechanisms of change were impossible, or even that it was absurd to imagine that they could be included.

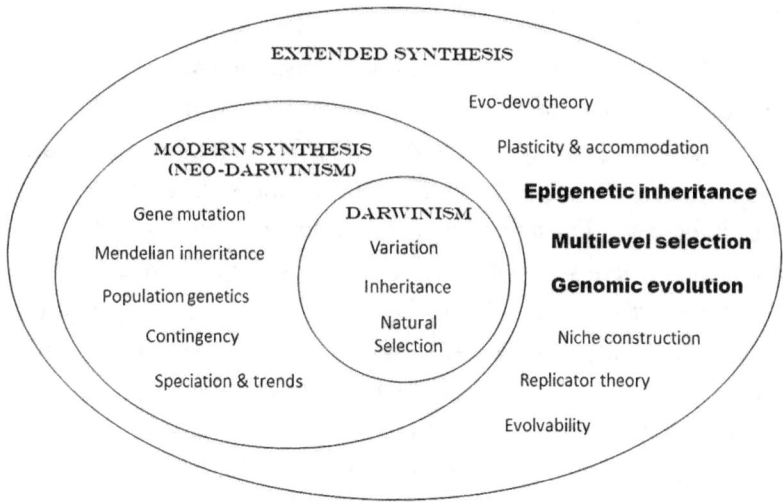

Figure 6.1: *How the EES relates to Darwinism and neo-Darwinism. The developments from Darwinism, to neo-Darwinism, and to the EES are all viewed as extensions of each other. The items in bold font are ones that seem to me to be incompatible with neo-Darwinism as shown in the next diagram (based originally on. (Based on a figure in Massimo Pigliucci and Gerd Müller, eds.,* Evolution: The Extended Synthesis. *Cambridge, MA: MIT Press, 2010.)*

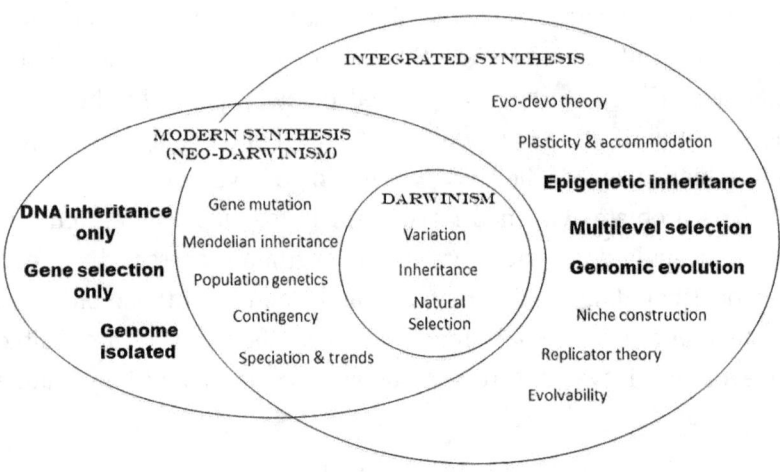

Figure 6.2: *The relations between Darwinism and later developments, rearranged to show how neo-Darwinism included ideas that have been falsified (bold face). These include the ideas: (1) that DNA alone specifies the organism and ensures its inheritance; (2) that selection operates only on genes as DNA (but that is also where some of the problems lie—see the diagram in my Major Statement explaining the different definitions of a gene); and (3) that the genome is isolated—based on the Weismann Barrier and on the so-called "Central Dogma of Molecular Biology." (Based on a figure in D. Noble, "Evolution Beyond neo-Darwinism: A New Conceptual Framework,"* Journal of Experimental Biology, *2015, 218: 7–13.)*

As I have already shown above, that process began very early in Weismann's 1893 writings. Essentially, his idea was that blind chance producing genetic variation followed by natural selection to winnow out the failures was not only possible, but also entirely sufficient. Darwin clearly did not believe that.[14] Darwin not only accepted

---

[14] Darwin wrote in the first edition of *On the Origin of Species*: "I am convinced that Natural Selection has been the main but not exclusive means of modification" (London: John

Lamarck's view that the inheritance of acquired characteristics had played a role, he actually formulated a physical theory (his theory of "gemmules") for how it could happen that changes in the soma could influence the germline. Recent research has shown that RNAs and imposed genome marking can serve precisely this role.

It seems to me therefore to be one of the original cornerstones of neo-Darwinism to insist that the inheritance of acquired characteristics, via epigenetic and other mechanisms, is impossible. There is now ample experimental evidence that it is possible, and that such transgenerational epigenetic changes can, in some cases, be transmitted down many generations. I dealt with this question in item 7 of my Major Statement. Much more detail and the full references can be found on the relevant pages of my website[15] and in *Dance to the Tune of Life* (Oxford UP, 2017).

There is also ample evidence that the genome is not isolated. As Barbara McClintock wrote when receiving her Nobel Prize in 1983, the "genome…is a highly sensitive organ of the cell."[16] James Shapiro has elaborated and extensively documented the evidence for genome re-organization (natural genetic engineering) in response to environmental influences in his book *Evolution, a View from the 21st Century*.[17]

I conclude, therefore, both that the original formulation of neo-Darwinism was falsifiable and that it has been falsified—if it is regarded, as many neo-Darwinists have insisted, as entirely sufficient.

Of course, that does not mean that the neo-Darwinist mechanism (chance genetic variations followed by natural selection) does not happen. Nor does it mean that all the impressive developments of population genetics using the mathematical ideas developed by neo-

Murray, 1859; p. 6). He reiterated this statement with increased force in the sixth edition published in 1872.

[15] [http://www.musicoflife.website/Answers-menu.html.]

[16] B. McClintock, "Responses of the Genome to Challenge," *Science,* 1984, **226**: 792–801.

[17] J.A. Shapiro, *Evolution: A View from the 21st Century.* New York: FT Press, 2011.

Darwinism suddenly become incorrect. That is no more so than a claim that quantum mechanics means that Newton's equations are no longer useful. It simply means, as Charles Darwin thought, that neo-Darwinism is not the only mechanism by which evolution may have occurred. My appeal is for a return to a more nuanced, multi-process view of evolutionary change. Darwin could not have anticipated what we have discovered today, but in his caution—and even in his specific theory of gemmules—he was far-sighted. If we return to a position that more closely resembles Darwinism by including the inheritance of acquired characteristics, then the term "neo-Darwinism" becomes redundant.

6.4. Why the Angst?

Why does this seemingly obvious, even innocuous, idea worry so many orthodox neo-Darwinists? So much so that they pour scorn on scientists like me who document the evidence that neo-Darwinism as a scientifically falsifiable theory is not the only show in town. This is a subject for professional historians and sociologists, of course, but I can at least venture a plausible reason—which is what I do in my new book.

The reason was first pointed out to me many years ago at a 1998 Novartis Foundation Symposium on the *Limits of Reductionism in Biology*,[18] when a distinguished neuroscientist told me quietly during a coffee break that he would go along with what I was proposing as an integrative view to complement reductionism, "except that it would let God back in."

It is undeniable that many dogmatic neo-Darwinists are also militant atheists and that they have used their conclusions from their interpretations of evolutionary biology to buttress their position. I am neither a theologian, nor am I conventionally religious. But I also know that many scientists who are conventionally religious

---

[18] Gregory R. Bock and Jamie A. Goode, eds., *The Limits of Reductionism in Biology* (Novartis Foundation Symposium 213). New York: Wiley, 1998.

and hold theistic views would laugh at some of the caricatures of their views presented by popularizing, atheistic neo-Darwinists. The theologians I know are often better philosophers than their critics and are far removed from the naïve fundamentalist views of many who use their religious beliefs to oppose the idea that life evolved. The problem is that the continuing and often acrimonious debate between neo-Darwinists and their fundamentalist opponents has polarized the debate, particularly so in the United States, to the point at which to back down even marginally on either side would involve great loss of face.

It is in that context that it is possible to understand why it seems necessary to militant atheistic scientists that they should exclude any idea that neo-Darwinism has been shown to be not the only possible mechanism of evolutionary change. That is why some of us who do dissent from the dogmatic forms of neo-Darwinism have formed the Third Way of Evolution website.[19] A steadily growing number of important scientists involved in work on evolutionary biology have joined us. One of my reasons for helping to launch this initiative is that many of the recordings of my own lectures to standard international congresses have been subsequently copied on, or embedded in, websites that are clearly run by creationists or those supporting Intelligent Design theories. I also know that some of my neo-Darwinist critics are quoting from those websites, not from the lectures or publications themselves. They are making the mistake of failing to consult the original sources. Anyone who can write that I proposed that "Darwin was rong" (*sic*) has simply not even bothered to consult what I said or wrote, particularly because I argue that he was largely right. Sadly, this is typical of the level of debate frequently found on the Internet.

---

[19] [https://www.thethirdwayofevolution.com/.]

6.5. Back to Tinbergen's Four Questions

It is therefore a pleasure to take part in a Dialogue with David Sloan Wilson. This is how debate in science should be done and, like him, I am also grateful to AcademicInfluence.com for the opportunity.

As I have already noted, Professor Wilson's definition of neo-Darwinism seems to me to be more like a framework for research —even a necessary framework—than a testable theory of evolution. My reading of Tinbergen's 1963 article also convinces me that this is the way he viewed his four questions. In his article, he doesn't point to a way in which his approach could be falsified. Questions, anyway, don't have the function to be falsified, unless we regard discovering conceptual mistakes embedded in a question to be a form of falsification. For example, a question like "Where is the West Pole?" cannot be interpreted in the same way as "Where is the North Pole?" It suffices to point out that no one knows what a West Pole would be like—or how we would know we had found it—to realize that it is not a valid question. But we know how to find the North and South Poles on a spinning object like the earth. "Where is the West Pole?" might well function in literature or poetry, but it doesn't have meaning in science. It is simply an ill-formed and useless question.

This short digression on the interpretation of questions is important because it is true to say that Weismann did come to the conclusion that his theory was necessarily true. But a valid empirical theory can't be necessarily true. Only theories in logic and mathematics can be necessarily true in the sense that they can be proved to be so. Any theory with empirical consequences must run the risk of being falsified if the world turns out not to behave in the way we supposed. I therefore believe that Weismann overstepped himself when he made that claim in his debate with Herbert Spencer. I feel sure that he must have thought he was proposing an empirically valid scientific theory. Why else would he have performed his tail-cutting experiments? The problem was that he felt certain he was right and translated that certain belief into his statement about its necessity. We have probably all made that mistake from time to time. It is sufficient to find that we cannot even imagine an exception to our certain belief, to slide into thinking that we have found a necessary empirical truth. Nineteenth-century determinists made a similar mistake when they thought that the world was necessarily determinate and they couldn't possibly have envisaged the arrival of quantum mechanics.

Tinbergen's questions seem to me to be a perfectly valid framework for asking the right questions. They are a kind of checklist to tell us whether we really have fully addressed what we need to study about an evolutionary biology problem. As Professor Wilson also says, there is still much research to be done within the framework of those four questions. In that sense, I understand his view that they are sufficient—even that they are necessary.

But I don't go so far as to regard any other questions as distractions. It does not seem to me to be a distraction to ask whether acquired characteristics can be inherited or whether genomes can be reorganized in response to environmental stress. Those are valid empirical questions and we now have the means for answering them. Moreover, the answers are often interesting and important. They lead to practical conclusions in a variety of fields, ranging from immunology (antigen invasion triggers targeted—and therefore functional —genomic change in immune system cells) to evolutionary biology (the same process occurring intergenerationally via natural genetic engineering) and the growing resistance of bacteria to antibiotics (they behave like the immune system in response to invasive stress).

I will illustrate the statement that acquired characteristics can be inherited with an example that includes an important aspect of animal behavior.

The example concerns a mechanism by which evolution can use epigenetic marking to bypass the Weismann Barrier by transmitting the epigenetic marks to offspring through behavior. This process has been demonstrated by Michael Meany's group in Canada. Rodents, like many other animals, groom their young by licking and stroking them. This behavior enhances the health and longevity of the progeny. It also influences epigenetic marking in the region of the brain called the hippocampus which, among other roles, plays a part in emotional behavior. The epigenetic effects can therefore predispose the progeny to show the same behavior towards their young. This form of epigenetic inheritance doesn't even require transmission through the germ line. It is a behavioral way of bypassing the Weismann Barrier.[20]

It seems to me, first, that it was a valid question to ask how this kind of behavior could be transmitted through many generations. In that sense, it was also a question well-formulated within Tinbergen's framework. If that is what Wilson means by "sufficient," then I suspect that almost any question about evolution can be interpreted as being within the framework, and it then seems trivially true that we don't need to go outside the framework. We then arrive at the situation in which one can ask a question within the Tinbergen framework forming Wilson's definition of neo-Darwinism that leads to a test of a different, and I would argue more usual, definition of neo-Darwinism. For the examples I have given show that acquired characteristics can be robustly inherited, and that in some cases this happens independently of transmission through the germline, contrary to the original formulation of neo-Darwinist theory. There are also examples of epigenetic inheritance that do go through the germline, either as genomic markings, or as accompanying RNAs.[21] These also break the Weismann Barrier.

Now, compare the results of such experiments with Weismann's statement:[22]

---

[20] I.C.G. Weaver, "Life at the Interface Between a Dynamic Environment and a Fixed Genome," in D. Janigro, ed., *Mammalian Brain Development.* New York: Humana Press, 2009; pp. 17–40.

[21] D. Noble, "How Widespread is Trans-generational Inheritance of Acquired Phenotypic Characteristics?," Denis Noble website. [https://www.denisnoble.com/answers-transgenerational-inheritance/.]

[22] A. Weismann, *Das Keimplasma: Eine Theorie der Vererbung.* Jena: Fischer, 1892. (English translation by Denis Noble.—Ed.)

When these deviations only affect the soma, they give rise to temporary non-hereditary variations; but when they occur in the germ-plasm, they are transmitted to the next generation and cause corresponding hereditary variations in the body.

If we take Weismann seriously, Meany's research clearly disproves his theory, as do a growing number of other examples of the trans-generational inheritance of acquired characteristics that I reference fully in *Dance to the Tune of Life*. Incidentally, these examples also demonstrate that Weismann's theory was indeed an empirically testable theory. It was also the foundation stone of neo-Darwinism.

Finally, before summing up my Response, I would like to tidy up a few detailed points which I find confusing. The first two are quotes from Massimo Pigliucci.[23]

...the [Modern Synthesis] was not something radically different from Darwinism and neo-Darwinism (the late 19th century modification of the original theory that got rid of Lamarckian influences, largely thanks to the work of August Weismann and Alfred Wallace).

This seems to me to be a very strange argument. Neo-Darwinism is significantly different from Darwinism precisely because it "got rid of Lamarckian influences." That is one of the main reasons I take issue with it. If excluding "Lamarckian influences" means excluding the inheritance of acquired characteristics, then neo-Darwinism was wrong to exclude them—particularly when Darwin himself did not.

This quotation, also from Pigliucci, confirms my view that the exclusion of Lamarckian influences was the main motive and corner-stone of neo-Darwinism:[24]

We are, however, in need of explicitly and organically incorporating into the framework of the MS a number of new discoveries and con-

---

[23] Massimo Pigliucci and David Sloan Wilson, "The Origin of the Extended Evolutionary Synthesis: An Interview with Massimo Pigliucci," This View of Life website, 2016. [https://thisviewoflife.com/the-origin-of-the-extended-evolutionary-synthesis-an-int erview-with-massimo-pigliucci/.]

[24] *Ibid.*

cepts (phenotypic plasticity, epigenetic inheritance, evolvability, and so forth) that were unknown to, or unappreciated by, the architects of the Modern Synthesis.

The problem here is that epigenetic inheritance is precisely the means by which forms of the inheritance of acquired characteristics become restored to evolutionary biology. So also is phenotypic plasticity, as Waddington showed when he first defined the concept of epigenetics. He also demonstrated ways in which epigenetic changes involving plasticity could become assimilated into genetic inheritance after a few generations. Since the exclusion of the inheritance of acquired characteristics was the main criterion by which Weismann and Wallace distinguished neo-Darwinism from Darwinism, to bring it back is to go back to Darwinism. Neo-Darwinism as defined by Weismann and Wallace is then falsified on its central idea.

It doesn't matter much whether we talk of "paradigm shifts" or not, but it does matter a lot whether neo-Darwinism remains clearly distinct from Darwinism. If we assimilate precisely what Weismann and Wallace intended to exclude, then I don't see why we continue to call it "neo-Darwinism." It reverts to being simply "Darwinism." This was, after all, why Waddington took issue with neo-Darwinism.

Next, consider this passage from Wilson's opening Statement:

> Complex interactions can produce lots of pattern at the system level, but they are no more likely than a point mutation to produce adaptive pattern without a process of selection.

I think this misses an important point about genome reorganization. Shuffling whole domains of sequences, which code for already-functional protein domains, can clearly improve the chances of new functional patterns appearing. To be sure, natural selection can then work on the outcome. But the source of genetic variation is no longer completely blind. If organisms can do this in response to environmental stress, then the process is no more blind than is the response of the immune cells to new antigens. Organisms can recruit stochasticity at the genome level and at the level of gene expression to greatly improve their chances of meeting new challenges.

I also find it hard to reconcile Wilson's position here with the following later quote from his Statement:

> Anything that smelled of directed evolution was branded with the label "Lamarckian" and declared impossible, much as group selection was declared impossible during the 1960s and 70s.
>
> Backing away from this dogmatic position is an important part of the extended evolutionary synthesis. For me, the best way to think about directed evolution is to focus on animal behavior, which brings the arc of this essay back to Tinbergen. In the conventional view of natural selection, mutations are not directed but they result in the evolution of behaviors that are indubitably directed....
>
> Since the work of James Mark Baldwin, we have known that directed behaviors that are a product of undirected evolution can double back to influence the evolutionary process. What the organism chooses to do by learning alters the selection pressures operating on the genes of the organism. This was celebrated as a major insight at the time—a form of directed evolution that was fully consistent with Weismann's doctrine....

I like the backing away from "this dogmatic position," but I am at a loss to reconcile these statements with the earlier ones. It might be best for me to pose my concern as a series of questions, since I may have misunderstood what is intended.

1. How can "a form of directed evolution" be consistent with "Weismann's doctrine?" *I thought the whole point of Weismann's work was precisely that evolution is not directed, but rather entirely due to blind chance followed by natural selection.*

2. And how can a form of directed evolution be consistent with the earlier statement that evolution has no foresight? *Surely, if we allow evolution to be guided by directed behavior that can*

> *"double back to influence the evolutionary process," then it*
> *no longer lacks foresight.*

3. What is meant by saying: "What the organism chooses to do by learning alters the selection pressures operating on the genes of the organism"? *If organisms can choose to alter selection pressure, then they are changing the direction of evolution in a functional way.*

Next, consider this:

Open forms of phenotypic plasticity are properly regarded as evolutionary processes built by other evolutionary processes, or "Darwin machines" to use the felicitous phrase coined by William Calvin and elaborated upon by Henry Plotkin. Unlike genetic evolution, Darwin Machines are expected to be directed forms of evolution, although they must also have an arbitrary component to remain open-ended. When Darwin Machines become transgenerational through forms of social learning found in many species and forms of symbolic thought that are distinctively human, there should be no stigma whatsoever about the fact that they are partially directed.

Again, I find it hard reconcile this quotation with the statement that evolution "lacks foresight."

And finally:

In the same way, ideas about directed evolution that have been worked out for the study of behavior are applicable to traits that are not customarily regarded as behavioral, such as genetic and epigenetic inheritance mechanisms, the immune system, developmental programs, and neural processes in the brain. Rigid adherence to Weismann's doctrine should be declared thoroughly obsolete, along with rigid rejection of group selection. These are profound advances in evolutionary thought, but they do not require going beyond Tinbergen's four questions. If anything, they require a proliferation of Tinbergen's four questions for every evolutionary process that is built by another evolutionary process.

I agree that these are "profound advances" and that "Rigid adherence to Weismann's doctrine should be declared thoroughly obsolete, along with rigid rejection of group selection." In fact, these are among the primary reasons for which I think that neo-Darwinism as originally formulated has been shown to be insufficient.

By the usual definitions of neo-Darwinism, I think we can safely say that the original theory has been falsified. Moreover, it seems to me that Professor Wilson and I would be in fairly good agreement on the reasons why that is so.

I agree that this doesn't change anything in relation to Tinbergen's four questions, but then I also do not think those questions can, in themselves, constitute an empirically testable theory.

Actually, unless I have seriously misunderstood these quotations, my questions on them are rather more than nit-picking. The more I think about them, the more they seem important and the more they seem to support my case.

6.6. Conclusions

So where does this leave Professor Wilson and me in this Dialogue?

My main conclusion is that Wilson and I seem to agree on many of the new trends in evolutionary biology and their significance, but we are working with fundamentally different definitions of neo-Darwinism. He seems to view neo-Darwinism as a *framework* rather than an empirically testable *hypothesis*. In that sense, it is more in the nature of metatheory—a theory about the kinds of questions we should be asking—rather than a hypothesis about the answers to those questions in any particular case.

Metatheory is important and scientists often neglect it. We all need frameworks in our research within which to formulate specific hypotheses. So, I agree that Tinbergen's four questions are important and that, as questions to guide research, they are sufficient. But I don't recognize the questions themselves to be a theory of evolution, and therefore not what neo-Darwinism has been about.

I have encountered some other scientists who believe that neo-Darwinism is more like a framework, and I also believe that this is why they feel they can accept many or even all of the new trends in evolutionary biology without thinking that neo-Darwinism is anything other than all-sufficient ("*Allmacht*," to quote Weismann again). But the "necessity" in Weismann's case cannot be the same as in Wilson's. Weismann surely believed he had formulated an empirical truth about the world (but mistakenly represented it as a necessary truth). I don't think Tinbergen's questions do that. They are not empirical propositions, even though answering them in specific cases could generate empirical statements. But those answers would agree with, or disprove, any hypotheses we may have *about those answers*, not the general framework questions themselves.

I think we can leave it to the historians of science to work out how neo-Darwinism sometimes seems to morph into metatheory, whereas its founding work was a very strong empirical hypothesis.

7. David Sloan Wilson's Final Reply

In the final installment of our Dialogue on Evolution, I will attempt to summarize our two positions and their implications in a stand-alone fashion. As those who have read the previous installments know, Professor Noble and I do not regard each other as adversaries and approached the question, "Is Neo-Darwinism Enough?," in different ways. While our Dialogue might lack the excitement of a boxing match, it has also been more productive, at least for me and hopefully for others, as well.

7.1. Neo-Darwinism as Meta-Theory

I took a classic article by Niko Tinbergen[1] as my conception of neo-Darwinism. Tinbergen observed that four questions need to be asked for all products of evolution, concerning their function, mechanism, history, and development. Tinbergen's four-fold distinction maps roughly onto a two-fold distinction made independently by Ernst Mayr[2] between ultimate causation (= Tinbergen's "function" and "history") and proximate causation (= Tinbergen's "mechanism" and "development"). I defended the position that Tinbergen's four

---

[1] N. Tinbergen, "On Aims and Methods of Ethology," *Zeitschrift für Tierpsychologie*, 1963, **20**: 410–433.

[2] E. Mayr, "Cause and Effect in Biology," *Science*, 1961, **134**: 1501–1506. (Reprinted in *idem, Evolution and the Diversity of Life: Selected Essays*. Cambridge, MA: Harvard University Press, 1976; pp. 359–371.)

questions are enough for the modern study of evolution and that to claim otherwise is a distraction.

Professor Noble agrees with my assessment, with the proviso that Tinbergen's four questions should be regarded as a meta-theory rather than a specific hypothesis. I cannot improve upon his own wording (from the conclusion to his Response):

"[Wilson] seems to view neo-Darwinism as a framework rather than an empirically testable hypothesis. In that sense it is more in the nature of a meta-theory, a theory about the kinds of questions we should be asking, rather than a hypothesis about the answers to those questions in any particular case. Meta-theory is important and scientists often neglect it. We all need frameworks in research within which we formulate specific hypotheses. So, I agree that Tinbergen's four questions are important and, as questions to guide research, they are sufficient."

I am happy to call Tinbergen's four questions a meta-theory, which means that Professor Noble and I have reached agreement on my side of the Dialogue. However, it remains to discuss what it means to be a meta-theory and to avoid the facile conclusion that all meta-theories are created equal.

Consider the meta-theory that questions concerning proximate causation are enough and questions concerning ultimate causation are not needed. Or consider the meta-theory that all four of Tinbergen's four questions are better asked on the basis of intelligent design rather than Darwin's theory of evolution. It is a testable hypothesis that these two meta-theories are objectively worse than a meta-theory based on Darwin's theory of evolution that asks all four of Tinbergen's questions. The hypothesis can be tested on the basis of the specific hypotheses that flow from each meta-theory and the proportion that prove to be correct.

These are not idle questions. Many—even most—scholars and scientists outside the biological sciences are not guided by Tinbergen's fully rounded, four-question approach, as I described at more length in my Statement and Response. Even within the biological sciences, questions about proximate causation frequently crowd out questions about ultimate causation. As a meta-theory, orthodox economics is as detached from reality as many religions. Hence, the single most important advance in evolutionary theory is arguably to get more people to employ Tinbergen's four-question approach.

7.2. Neo-Darwinism as Specific Hypotheses

What Professor Noble calls a "meta-theory" bears a family resemblance to what Thomas Kuhn called a "paradigm" and Imre Lakatos called a "research program." None of these concepts have crisp definitions, but they all refer to configurations of ideas that cannot be falsified by a single experiment, the way that more specific hypotheses can within a given meta-theory, paradigm, or research program. Most practicing evolutionary scientists, historians of science, and philosophers of science would call neo-Darwinism closer to a meta-theory, paradigm, or research program than one or a few testable hypotheses. Nevertheless, I am happy to follow Professor Noble down this path for purposes of argument.

Professor Noble is especially concerned with the hypothesis that variation is blind with respect to what is selected, so that evolution is not directed in any sense. I am happy to agree with him that this hypothesis has been falsified and I provided examples of my own in my previous installments, so we have reached agreement on his side of the Dialogue. As with the concept of meta-theories, however, it is important to be clear on what it means for evolution to have a directed component and to avoid the facile conclusion that all hypotheses about directed evolution are created equal.

In the first place, it is important to keep in mind — as Professor Noble is also careful to stress—that a great deal of variation is blind with respect to what is selected. For example, it is remarkable how well Richard Lenski's long-term experiments on evolution in *E. coli*, now exceeding 60,000 generations, can be explained on the basis of blind variation without needing to invoke any kind of directed evolution. (Go here for my interview with Lenski, which is framed in terms of Tinbergen's four questions.)

In the second place, all mechanisms that result in directed evolution are descended from undirected evolutionary processes, as the social psychologist Donald T. Campbell[3] insisted with his phrase "blind variation and selective retention." Not only is this empirically the case to the best of our current knowledge, but there are also strong theoretical reasons to expect it to be the case.

In the third place, many examples of directed evolution, such as trans-generational epigenetic inheritance, are extremely limited in their potential for long-term, open-ended evolutionary change.[4] The distinction between closed and open forms of phenotypic plasticity is important in this regard. In closed phenotypic plasticity, the organism has evolved a fixed repertoire of traits (e.g., a fast or slow life history strategy) and environmental cues determine which trait is expressed. Traditionally, closed phenotypic plasticity was thought to be intra-generational, which means that the trait expressed by the parent does not influence the trait expressed by its offspring. Now we know that some examples of closed phenotypic plasticity are trans-generational, such as the example of mother rats influencing the life history strategies of their offspring by licking them to different degrees, which Professor Noble describes in his Response. Examples such as this are new and wonderful, but because they remain forms of closed phenotypic plasticity, they have little potential for long-term, directed evolutionary change.

Now let's consider open forms of phenotypic plasticity, such as operant conditioning made famous by B.F. Skinner.[5] Imagine placing a rat in a Skinner Box. The rat eventually presses a lever in its explo-

---

[3] D.T. Campbell, "Blind Variation and Selective Retention in Creative Thought as in Other Knowledge Processes," *Psychological Review*, 1960, **67**: 380–400.

[4] E. Jablonka and M.J. Lamb, *Evolution in Four Dimensions: Genetic, Epigenetic, Behavioral, and Symbolic Variation in the History of Life*. Cambridge, MA: Bradford Books/MIT Press, 2006. (Revised ed., 2014.)

[5] B.F. Skinner, "Selection by Consequences," *Science*, 1981, **213**: 501–504.

ration of the box, causing it to receive a food pellet. This rewarding experience causes the rat to press the lever again at a higher frequency than whatever else it was doing while exploring the box. After a few repetitions of the experience, the rat spends most of its time pressing the lever. The rat's behavior has been "selected by consequences," as Skinner put it, in the same way that genes are selected by their consequences. One of Skinner's most important insights was to make this connection between open-ended learning and open-ended genetic evolution. Essentially, it meant that Tinbergen's four questions can be asked for an individual as an evolutionary system in its own right.

To see what open-ended learning means for long-term directed evolution, consider the documented case of a coastal population of Japanese macaques that were provisioned with potatoes thrown onto a beach (this and other examples are discussed by Avital and Jablonka[6]). Eventually, one of macaques learned to wade into the sea to wash the sand off the potatoes and the learned behavior spread to the other macaques. The fact that most of the macaques did not learn to do this on their own and were slow to learn it from others is interesting and important but can be set aside for the moment. Now the whole population of macaques was spending a lot more time wading in the sea than they were before, just as the rat in a Skinner box spends a lot more time pressing the lever. This is likely to be hugely consequential for subsequent genetic evolution. We can imagine the macaques discovering other food resources, such as shellfish, which they never would have encountered before. As they deplete the shellfish close to the shore, we can imagine them wading further out. Since these learned activities are consequential for survival and reproduction, they would direct the genetic evolution of relevant traits such as the ability to swim and to see underwater. One can well imagine that marine mammals evolved by such a process directed by

---

[6] E. Avital and E. Jablonka, *Animal Traditions: Behavioural Inheritance in Evolution.* Cambridge: Cambridge University Press, 2001.

learning. (An "aquatic ape" hypothesis has even been proposed for human evolution.[7])

I love this scenario for genetic evolution directed by learning, but it is still also notable for its haphazard quality. Genetic evolution is directed by what is immediately rewarding without any regard to long-term consequences. Most of the macaques couldn't even foresee the advantages of washing potatoes when other macaques were doing it in front of their faces. Also, the example does not eliminate trial-and-error evolution, but merely relocates it to trial-and-error learning. It's not obvious how the "genetic evolution directed by learning" scenario alters expectations about long-term evolution of a species or adaptive radiations. A neo-Darwinian could also have predicted that coastal populations of monkeys are likely to evolve to use marine resources. Perhaps this is why the idea of genetic evolution directed by learning was regarded as important when proposed by James Mark Baldwin early in the twentieth century, but nevertheless didn't seem to go anywhere.

To find examples of evolution directed by more foresight, we must turn to human gene-culture co-evolution. I will end this essay by reviving an account given by the French Jesuit priest and paleontologist Pierre Teilhard de Chardin in his book *The Phenomenon of Man.*[8] At the time, Teilhard was a respected scientist and the introduction to his book was written by Julian Huxley, one of the architects of the Modern Synthesis. Today, Teilhard is forgotten as a scientist and his book is read mostly for its spiritual quality. Nevertheless, he was remarkably prescient and requires little updating from a modern evolutionary perspective, as I will now show.

Unlike some authors who argue that human-like intelligence is an inevitable outcome of evolution, Teilhard described human evolution as a happy accident, a combination of traits that just happened to come together (Stephen Jay could also defended this position with

---

[7] E. Morgan, *The Aquatic Ape Hypothesis.* New York: Stein & Day Publishers, 1982.

[8] P. Teilhard de Chardin, *The Phenomenon of Man.* New York: Harper & Brothers, 1959. (Originally published as *Le phénomène humain.* Paris: Éditions du Seuil, 1955.)

vigor). To the best of our current knowledge, human evolution is an example of a major evolutionary transition, which is indeed a rare event in this history of life. A major transition occurs when mechanisms evolve that suppress the potential for disruptive within-group selection, so that between-group selection becomes the main evolutionary force and groups become so cooperative that they become a higher-level organism in their own right (see Wilson [2015] and my Interview, Statement, and Response, for a more detailed account).[9] Other examples of major transitions include the first cells, nucleated cells, multi-cellular organisms, eusocial insect colonies, and possibly the origin of life itself as groups of cooperative molecular interactions.

Even something as rare as a major transition is only necessary and not sufficient for human-like foresight. In addition, cooperation within groups needs to take the form of transmitting large amounts of learned information across generations. Even naked mole rats, another mammal species that qualifies as a product of a major transition, do not cooperate with each other in this way.

Once hominids evolved the ability to rapidly adapt to their environments, they spread over the globe, occupying all climatic zones and dozens of ecological niches, while remaining tribes of a few thousand people subdivided into smaller groups. Teilhard describes this eloquently by asking us to imagine the tree of life growing slowly. Suddenly, one of the twigs on one of the branches starts to grow rapidly, until it overtops the rest of the tree. That description accurately describes the human cultural adaptive radiation, for better or for worse.

The kind of directed gene-culture co-evolution that took place during this phase of human evolution was little different from the scenario that I just described for Japanese macaques. It was directed toward short-term gain without regard to long-term consequences.

---

[9] D.S. Wilson, *Does Altruism Exist? Culture, Genes, and the Welfare of Others.* New Haven, CT: Yale University Press, 2015.

And the long-term consequences were far more likely to be negative than positive, such as the extinction of prey species, the degradation of the physical environment, and chronic inter-group conflict.

A recent example of cultural evolution that is directed at a small spatial and temporal scale and becomes undirected at a larger scale is provided by the Nuer, a pastoral African tribe that was in the process of replacing an adjacent tribe called the Dinka when contacted by Europeans in the nineteenth century.[10] Both tribes inhabited the same physical environment and had the same subsistence practices. The Nuer were historically derived from the Dinka as a lineage that became sufficiently distinct to acquire its own tribal identity. Both the Nuer and Dinka were highly intentional in how they lived their lives: growing millet, tending their cattle, raising their families, and negotiating their social relationships. The fact that the Nuer were slowly nibbling away at the Dinka territory was almost entirely the result of unforeseen consequences of lower-level intentional practices. There is no evidence that anyone from either tribe reflected upon what was happening at the tribal level or what they might do about it. Lower-level intentional processes might as well have been random as far as their higher-level consequences were concerned!

Eventually, human cultural evolution led to a positive feedback cycle in which the ability to produce food led to larger social groups, which in turn enhanced the ability to produce food, leading to the mega-societies of today. This period of human evolution is admirably described by Peter Turchin in his book *Ultrasociety: How 10,000 Years of War Made Humans the Greatest Cooperators on Earth*.[11] Teilhard eloquently asks us to imagine "tiny grains of thought" that gradually coalesce into larger and larger entities. Regretfully, however, directed cultural evolution in larger human groups is still dominated by the pursuit of short-term rewards without regard to

---

[10] R.C. Kelly, *The Nuer Conquest*. Ann Arbor, MI: University of Michigan Press, 1985.

[11] P. Turchin, *Ultrasociety: How 10,000 Years of War Made Humans the Greatest Cooperators on Earth*. Storrs, CT: Baresta Books, 2015.

long-term consequences, which more often than not are harmful to the common good.

Teilhard optimistically predicted that the coalescing process would eventually result in a single global consciousness that he called the Omega Point, which would regulate life on earth like a single organism. It goes without saying that humanity is very far from realizing that goal. Currently, we collectively have the same degree of foresight as cancer cells evolving themselves to extinction. However, we can still ask the question: "Is it theoretically possible for humans to direct their evolution toward global cooperation—to steer toward the Omega point, so to speak?" In my opinion, the answer to this question is "yes," although it won't be easy.[12]

For the purpose of this Dialogue, we can conclude that the neo-Darwinian hypothesis about evolution being entirely undirected has been falsified, but that the consequences of directed evolution are highly complex and by no means entirely benign.

I thank AcademicInfluence.com for organizing this Dialogue and Professor Noble for serving as my partner. I think we can all agree that we are living in exciting times as far as the development of evolutionary theory is concerned.

---

[12] For further discussion, see here. [https://web.archive.org/web/20170508143728/https://evolution-institute.org/article/steering-toward-the-omega-point-a-roundtable-discussion-of-altruism-evolution-and-spirituality/.]

8. Denis Noble's Final Reply

This Dialogue has proved very interesting and I appreciate both the invitation from AcademicInfluence.com to take part, and the intelligent and courteous way in which it has been conducted, both by the moderator (who was also the interviewer) and by Professor Wilson. I also appreciate some boof the criticism of my position in the latter's Response and his suggestions for further reading.

In this final contribution I will not attempt to answer all the questions that arise. I will focus my remarks on what I see as the central issues on which to sum up my reactions to the Dialogue as a whole.

8.1. Can Neo-Darwinism Be Redefined as Tinbergen's "Four Questions"?

1. Neo-Darwinism as originally defined by Weismann and Wallace was a specific theory of evolution, capable of empirical falsification.

Neo-Darwinism as defined by its nineteenth-century founders, August Weismann and Alfred Russel Wallace, was an experimentally testable theory of evolution. Tinbergen's "four questions" do not express the specific empirical predictions of neo-Darwinism and, as questions, they are not empirically testable. If neo-Darwinism is to be redefined as Tinbergen's four questions, as Wilson proposes, then it would no longer be a *theory* of evolution and therefore would not be what neo-Darwinism was about when it was formulated.

That theory has been falsified. It also seems to me that Wilson agrees that this is the case, when he says in his Statement:

> Anything that smelled of directed evolution was branded with the label "Lamarckian" and declared impossible, much as group selection was declared impossible during the 1960s and '70s. Backing away from this dogmatic position is an important part of the Extended Evolutionary Synthesis.

2. What can questions do?

Questions might imply theories, and even be theories masquerading as questions, for example when the questions are rhetorical.[1] But Tinbergen's four questions are clearly not rhetorical. Nor do they form a specific theory of evolution. As I argued in my Response, instead they form a valid framework within which specific theories can be developed.

They are based on Aristotle's "four causes," on which I elaborate in *Dance to the Tune of Life*. I show there that the four causes require extension to include attractors, coding, and reasons,[2] but I don't see those extensions as fundamental. An attractor can be seen as an evolved final cause, coding can be seen as a formal cause, and reasons can also be seen as a final cause.

As a framework for research, I therefore agree with Wilson that Tinbergen's four questions are sufficient.

3. How does neo-Darwinism differ from Darwinism?

Neo-Darwinism was formulated by Weismann and Wallace explicitly to correct what they thought was the incorrect acceptance of the

---

[1] As an example, Richard Dawkins's question to Lynn Margulis in their 2009 debate is clearly rhetorical: "It [Neo-Darwinism] is highly plausible, it's economical, it is parsimonious, why on earth would you want to drag in symbiogenesis when it's such an unparsimonious, uneconomical [theory]?" Margulis's reply was simply, "Because it's there." (See D. Noble, *Dance to the Tune of Life: Biological Relativity*, Cambridge: Cambridge University Press, 2017; p 138.)

[2] D. Noble, *Dance to the Tune of Life: Biological Relativity*. Cambridge: Cambridge University Press, 2017; pp. 176–181.

inheritance of acquired characteristics in Darwin's work. If neo-Darwinism is now to be redefined as accepting the inheritance of acquired characteristics, then we no longer have the original reason for distinguishing it from Darwinism.

Modern theories of evolution can then be seen as deriving from Darwinism, rather than from neo-Darwinism. It seems to me that the time has come to drop the prefix "neo-." Following the lead of Conrad H. Waddington,[3] I am perfectly happy with describing my view of evolutionary biology as originating with Darwin (and with Lamarck), but I don't see it as originating with neo-Darwinism.

4. What do people generally understand by neo-Darwinism?

Professional philosophers of science, popularizers of science, and the public at large naturally and almost universally understand this to be the original definition of neo-Darwinism.

It is impossible to read, for example, the recent book *Mind and Cosmos*[4] by the distinguished philosopher Thomas Nagel as a philosophical attack on Tinbergen's four questions. He doesn't even refer to Tinbergen. The book is clearly a sustained and fundamental attack on original neo-Darwinist theory as formulated by Weismann and Wallace and subsequently developed into the Modern Synthesis.

Similarly, Mary Midgley's articles and books[5] criticizing neo-Darwinism, in response to *The Selfish Gene*, also clearly deal with the original theory, not Tinbergen's four questions, which are not even referenced.

As another example, John Dupré in his excellent book, *The Disorder of Things*,[6] is very critical of the neo-Darwinist perspective

---

[3] C.H. Waddington, *The Strategy of the Genes: A Discussion of Some Aspects of Theoretical Biology.* London: Routledge, 2014. (Originally published in 1957.—Ed.)

[4] T. Nagel, *Mind and Cosmos: Why the Materialist Neo-Darwinian Conception of Nature is Almost Certainly False.* Oxford: Oxford University Press, 2012.

[5] For example, M. Midgley, "Gene-Juggling," *Philosophy,* 1979, **54**: 439-458; and *idem, The Solitary Self: Darwin and the Selfish Gene.* Durham, U.K.: Acumen Publishing, Ltd., 2010.

[6] J. Dupré, *The Disorder of Things: Metaphysical Foundations of the Disunity of Science.* Cambridge, MA: Harvard University Press, 1993.

(although not by name), but I suspect that he would be happy enough with Tinbergen's four questions.

I also consulted Gregory Bateson's classic, *Steps to an Ecology of Mind,* in which we find the following:[7]

> Lamarck, probably the greatest biologist in history, turned that ladder of explanation (starting with the supreme mind, followed by man and then down to the infusoria) upside down. He was the man who said that it starts with the infusoria and that there were changes leading up to man. His turning the taxonomy upside down is one of the most astonishing feats that has ever occurred. It was the equivalent in biology of the Copernican revolution in astronomy.

This could not possibly have been written from a neo-Darwinist perspective. Lamarck was the *bête noire* of neo-Darwinism. Bateson knew of Tinbergen's work,[8] but does not refer to or criticize the four questions.

In fact, looking through my own collection of books and articles that discuss neo-Darwinism, I simply can't find any that criticize the four questions.

5. Does it matter?

I remain puzzled by what I see as a completely new definition of neo-Darwinism. I am therefore forced to conclude that Professor Wilson must see the original definition of neo-Darwinism as no longer tenable.

As he says himself in his Statement, the hypothesis of the Weismann Barrier has been demonstrated to be wrong:

"Rigid adherence to Weismann's doctrine should be declared thoroughly obsolete, along with rigid rejection of group selection."

The implications of this development are too important to be subsumed into a new definition carrying the name of "neo-Darwinist"

---

[7] G. Bateson, *Steps to an Ecology of Mind.* Chicago: University of Chicago Press, 1972; p. 433.

[8] *Ibid.*; p. 181.

theory since they contradict the original, and usual, version of that theory.

Moreover, they contradict it in ways that fundamentally change our view of the nature of living organisms. I enlarge on this point in section C.

8.2. Evolution Itself Evolves

The key difference between hard-line neo-Darwinism and the position that both Wilson and I favor, albeit in somewhat different ways, is whether the evolutionary process is directed.

Wilson writes in his Statement:

> Since James Mark Baldwin . . ., we have known that directed behaviors that are a product of undirected evolution can double back to influence the evolutionary process. What the organism chooses to do by learning alters the selection pressures operating on the genes of the organism. This was celebrated as a major insight at the time—a form of directed evolution that was fully consistent with Weismann's doctrine . . .

I agree with the majority of this statement, although, following Patrick Bateson,[9] I prefer to give the credit to Douglas Spalding (1841–1877), who had the idea before Baldwin.

As I have already explained in this Dialogue, the part of this quotation that I don't understand is the idea of consistency with Weismann's doctrine. I cannot see how directed evolution could be consistent with Weismann's ideas of everything being generated by blind chance.

There is a deep misunderstanding about the role of blind chance in evolution. As any good physicist will tell you, stochasticity at low (e.g., molecular) levels is perfectly consistent with order at higher levels, as for example described by thermodynamics. Everything, including organisms, must live with the existence of stochasticity at

---

[9] P. Bateson, "The Adaptability Driver: Links between Behavior and Evolution," *Biological Theory*, 2006, **1**: 342–345.

low levels. Organisms must therefore harness stochasticity to generate functionality.[10] It is functionality—and its ability to constrain what happens at lower levels—that is a necessary prerequisite of directed evolutionary change.

Blind chance was therefore necessarily involved in the evolution of processes that enable directed evolution. The evolution of cells in the immune system in response to challenges from new antigens illustrates one of the main processes involved. Faced with a new antigen challenge, the mutation rate in the variable part of the genome can be accelerated by as much as one million times. So far as we know, those rapid mutations occur randomly. If we focus our attention on just that part of the genome, the process will appear to be neo-Darwinian. But the location in the genome is certainly not random. If such a high rate of random mutations were to occur throughout the genome the organism would almost certainly not survive. Necessary functionality in the fixed parts of the proteins involved would be lost. The functionality in this case therefore lies precisely in the targeting of mutation at the relevant (variable) part of the genome and the preservation of the fixed part. The mechanism is directed, because the binding of the antigen to cell receptors in the secondary lymphoid organs (spleen and lymph nodes) itself activates the proliferation process. Organisms can therefore respond to environmental challenges by rapidly and selectively mutating just parts of their genomes. In so doing, they use targeted stochasticity to achieve the goal.

This example shows that the process can occur in the lifetime of an individual organism. There is no reason why this kind of mechanism should not be used in evolutionary change, and it is.

A well-known, functionally driven form of genome change is the response to starvation in bacteria. Starvation can increase the targeted reorganizations of the genome by five orders of magnitude, i.e., by

---

[10] This was the main point of my lecture at the Physiological Society on 21 November, 2016 (which was a longer version of the lecture given at the recent Discussion Meeting of the Royal Society and the British Academy, 7–9 November, 2016).

a factor of over 100,000. Again, the location of the mutations is targeted. This is one of the mechanisms by which bacteria can evolve very rapidly and in a functional way in response to environmental stress by triggering rapid mutation in selected parts of the genome without disturbing the rest.

These hypermutation mechanisms have been investigated by Richard Moxon and his colleagues who use the term "contingency locus" to characterize the targeted loci of hypermutable DNA.[11] Since "mutation rates vary significantly at different locations within the genome," they propose that "it is precisely in the details of these differences and how they are distributed that major contributions to fitness are determined." In an earlier article, Moxon and Thaler write:[12]

> This phenotypic variation, which is stochastic with respect to the timing of switching but has a programmed genomic location, allows a large repertoire of phenotypic solutions to be explored, while minimizing deleterious effects on fitness. (emphasis added)

I have emphasized the last phrase because it is the key to understanding such goal-directed processes in organisms. What is random (blind chance) when one considers the generation in time in the hypermutation regions is consistent with directed evolution when one considers the spatial targeting of the activation of hypermutation.

So, blind chance and natural selection may well have been sufficient to allow the emergence of directed hypermutation. But once it has done so, evolution itself evolves.[13] There is a kind of bootstrapping here. There is no going back. Organisms that have directed, functionally significant hypermutation will clearly win out and will pass that ability on to subsequent generations.

---

[11] R. Moxon, C. Bayliss, and D. Hood, "Bacterial Contingency Loci: The Role of Simple Sequence DNA Repeats in Bacterial Adaptation," *Annual Review of Genetics*, 2006, **40**: 307–33.

[12] E.R. Moxon and D.S. Thaler, "The Tinkerer's Evolving Tool-Box," *Nature*, 1997, **387**: 659–62.

[13] "Evolution Evolves" is the title of a focused issue of the *Journal of Physiology* devoted to physiology and evolution: *Journal of Physiology*, 2014, **592:** i–iv, 2237–2438.

Evolution is not therefore a single parsimonious process. There have been many such bootstrappings, each of which changes the subsequent evolutionary process by adding new directional mechanisms to the armory of mechanisms used by nature. The neo-Darwinist mechanism itself is also in this category. Before the evolution of multicellularity, the Weismann Barrier could not and did not need to exist. And when it did arise, nature found ways of circumventing it so that it is a relative rather than an absolute barrier.

So, was the watchmaker really blind? This is not now as easy a question as it may have seemed when Richard Dawkins wrote *The Blind Watchmaker*.[14] The mechanisms that underpin directed evolution have evolved. Having done so, the genie cannot be put back into the bottle. Perhaps we should say that the Watchmaker may have started off blind, but she has become one-eyed.[15] Subsequent evolution is not pure chance. The implications of that change in view are profound, as I lay out in the final section.

8.3. The Implications

The existence of directed evolution is the key. And I believe that both Wilson and I see it as such, although we may have different ways of spelling out the implications. He sees it as another extension of what he chooses to call neo-Darwinism.

My view is that there could not be a more important difference between the neo-Darwinist Modern Synthesis as usually defined and the new trends in evolutionary biology leading towards the acceptance of forms of directed evolution. Whether we see the world as governed by pure chance or at least partially by directed change is fundamental to the humanities and social sciences at least as much as (if not more than) to physics, chemistry, biology, and engineering.

---

[14] R. Dawkins, *The Blind Watchmaker: Why the Evidence of Evolution Reveals a Universe without Design*. New York: Norton, 1986.

[15] I am indebted to the philosopher Sir Anthony Kenny for suggesting this expression for treating the Blind Watchmaker issue in the light of new trends in evolutionary biology.

Many in the humanities and social sciences view the original and dogmatic forms of neo-Darwinism—restricting evolutionary change to blind chance and natural selection—with at least suspicion and sometimes with outright hostility. The humanities in particular find it much more conducive to think that there is genuine purpose. Mere belief or wishful thinking are not valid reasons of course for ignoring whatever scientific investigation tells us, if that really is that there is no purpose. But it is nevertheless obvious that a version of evolutionary biology that includes directed change is of more than casual interest. And if Wilson and I are right that there is directed evolution, then it is important to the humanities and social sciences to make this absolutely clear. It is a fundamental change in evolutionary biology.

Moreover, the social science and humanities implications do not really depend on whether directed evolution originally arose from purely chance events or whether it should be viewed as an inevitable feature of the universe.[16] Once directed evolution has emerged it takes over, just as an attractor takes over once it has formed in complex systems.

This property of attractors is not just a property of biological systems. The air molecules of a tornado obey the same physicochemical equations as other air molecules, but once they are drawn into the tornado, they also conform to the high-level dynamics of the tornado. So, to take an example at a vastly bigger scale, do the particles, planets, stars, and other components of a rotating spiral galaxy. Similarly, at a micro-scale, the channel proteins, membrane lipids, and charged ions in heart cells obey the standard laws of chemistry, but are also constrained to behave as a coordinated ensemble when generating the electrical rhythm of the heart.

---

[16] Some cosmologists consider that the fine tuning of the cosmological constants might suggest that much of what has emerged during the evolution of the universe and of life may be an inevitable consequence of the fine-tuning, however that came about. To avoid this conclusion some cosmologists propose the existence of multiple universes, or a "multiverse," but at this stage in our knowledge that can best be described as pure metaphysical speculation. It is hard to see how any empirical test of the "multiverse" hypothesis could be formulated, let alone performed.

How this can happen is explained by the mathematics of Biological Relativity.[17] There is nothing fuzzy or mystical about the forms of causation involved. They can be represented rigorously in mathematical form. For those who know enough mathematics to appreciate the point, the constraints of the components by the whole enter into the solutions of the differential and related equations by determining the initial and boundary conditions that are essential for there to be particular solutions of the equations.

If the reader finds any of these ideas interesting, even exciting, there is much more on the same theme in *Dance to the Tune of Life*.[18]

---

[17] D. Noble, *Dance to the Tune of Life: Biological Relativity*. Cambridge: Cambridge University Press, 2016; Chapter 6.

[18] *Ibid.*; Chapters 7–9.

9. David Sloan Wilson's Afterword

My extensive exchange with Denis Noble on the theme "Is Neo-Darwinism Enough?" was a fascinating personal experience. We are not opponents, but rather two veterans of the scientific wars surrounding the development of evolutionary theory. Denis is 85 and I am 71, so we personally experienced what the current generation can only read about.

To call scientific inquiry a war is an exaggeration in some respects but appropriate in other respects. Chess can accurately be described as a battle, for example, even though no blood is shed. Like chess, science often takes the form of a battle that is highly constrained by a set of rules so that it results in the positive outcome of knowledge generation. And the stakes are often high! It is a myth that the pursuit of scientific knowledge is dispassionate and value-free. Scientists are flesh-and-blood people with value-laden worldviews that they import into their theories, whether they are conscious of the fact or not. It is for this reason, as much as the complexity of the ideas being considered, that scientific controversies are often so protracted and some scientists go to their graves without changing their minds.

Like a major war such as World War II, the evolution war has taken place along multiple fronts. My Dialogue with Denis covers at least three major battle zones: (1) the integration of Mendelism and Darwinism, leading to the Modern Synthesis in the 1940s; (2) the spread of evolutionary thinking into historically separate disciplines such as ecology and ethology (the study of animal behavior), represented by Niko Tinbergen and his four questions, which took place roughly during the 1960s; and (3) the widespread rejection of multilevel selection, which also took place roughly during the 1960s.

A bit of work was required for Denis and me to decide that we were recounting different battles! Denis is primarily a veteran of the Modern Synthesis battle. On the positive side, this was a genuine synthesis of Mendelism and Darwinism, which previously had been regarded as opposing theories. On the negative side, it led to a portrayal of evolution as without purpose, fiercely denied the possibility of anything that could be called Lamarckian, and privileged genes over richly interconnected developmental and physiological systems that include genes, plus so much more. With this particular battle in mind, Denis wants to declare total victory when Lamarckian forms of inheritance have been identified, and I'm happy to go along with him on that.

My part of the Dialogue draws attention to the spread of evolutionary thinking into disciplines such as ecology and ethology. This was less a battle than a diffusion of a powerful set of ideas, but it still required time and encountered various obstacles. Niko Tinbergen, Konrad Lorenz, and Karl von Frisch did their pioneering work on ethology from an evolutionary perspective in the 1930s. Tinbergen's "four questions" article, which provided a concise account of how all products of evolution need to be studied, was published in 1963.[1] Independently, Ernst Mayr made the basic distinction between proximate and ultimate causation.[2] These have become part of the basic canon of evolutionary theory and my part of the Dialogue was to insist that they need to remain so.

Notice that these two controversies are largely orthogonal to each other. Tinbergen's four questions say nothing about genes *per se* —only *mechanism* as one of the four questions. If some of the mech-

---

[1] N. Tinbergen, "On Aims and Methods of Ethology," *Zeitschrift für Tierpsychologie*, 1963, **20**: 410–433.

[2] E. Mayr, "Cause and Effect in Biology," *Science*, 1961, **134**(3489): 1501–1506.

anisms of inheritance turn out to be Lamarckian, even resembling Darwin's speculations about gemmules, then that is interesting and important but doesn't overturn the four-question approach. In this sense, Denis is right to point out that the four-question approach is not a falsifiable hypothesis, but that doesn't detract from its importance as a way of thinking that needs to be widely understood and put to practical use.

The controversy over multilevel selection (MLS) is orthogonal to the other two. Richard Dawkins was a student of Niko Tinbergen and writes about the four-question approach as enthusiastically as I do. However, we part company on the importance of group selection. In some respects, group selection is a falsifiable hypothesis. If we compare fitness differences among individuals within groups and fitness differences among groups in a multi-group population, we can ask the question: "Is between-group selection ever a significant evolutionary force, compared to within-group selection?" The answer to that question is "Yes—not always, but in many cases." In that sense, I can declare total victory against those who maintain that group selection is virtually never a significant evolutionary force, just as Denis can with respect to Lamarckian mechanisms of inheritance.

But the clear-cut hypothesis testing requires defining key terms such as "selfishness," "altruism," "individual selection," and "group selection" in a certain way. Individual and group selection are selection within and between groups, respectively. Selfishness is what evolves by within-group selection. Altruism is what evolves by between-group selection. It turns out that these key terms can be defined in different ways, which also have intuitive appeal. Individual selection can be defined as "the fitness of individuals, averaged across all social contexts." Altruism can be defined as "a behavior that increases the fitness of others and results in a net loss in the altruist's fitness (as opposed to relative to other members of the same group). Formal models can be built around these definitions that come to the same conclusions about what evolves but assigns them to different categories, so that what counts as altruistic under one framework counts as selfish under another. One framework, Selfish Gene Theory, defines everything that evolves by genetic evolution as a form of selfishness because, after all, those genes were more fit than their alternatives, all things considered.

To get an intuitive feel for these different mental configurations, imagine that you were raised to be nice to others, to avoid selfish behaviors, and to steer clear of other selfish people. You are also empathic by nature and derive pleasure from helping others, especially when it is reciprocated. When you are asked to help others or support a good cause, you are likely to comply without asking what's in it for you. Then you meet someone in a bar who insists that everything— *everything*—is a form of self-interest. Your giving behavior can be explained by the fact that it makes you feel good, for example. This person does not necessarily behave badly. She might also be willing to help others and refrain from predatory behaviors, but only when conceptualized as a form of enlightened self-interest. You might or might not decide to befriend this person, but you will probably *never* alter the way that she views the world. You'll just have to translate everything she says into your own language: What's altruistic vs. selfish for you is enlightened vs. stupid forms of self-interest for her.

If you do befriend this person, you might discover that despite your ability to translate her language into yours, they are not equivalent in terms of how they cause the two of you to behave. Giving for you is more spontaneous, whereas for her it requires cognitive work to align it with her perceived self-interest. This means that you actually are the more giving person as far as your behaviors are concerned. That's nice for others, but it also sets you up for being taken advantage of, whereas your friend is less likely to get burned. You begin to realize that there are costs and benefits associated with both ways of thought, which might explain why they co-exist. You grew up in a nurturing household where "be nice" was a good rule of thumb. Your friend wasn't so lucky and needed to be constantly on the lookout for being taken advantage of. Also, you came from a Christian tradition where "do unto others" and original sin are part of the cultural DNA. Her father was an atheist and economist, where the assumption of individuals as self-interested "atoms" is part of the cultural DNA. Your worldviews might owe themselves to your cultural traditions, just as much as your individual circumstances.

I have told this story to make four important points. First, human worldviews in the real world are *diverse* at all scales. Pick any person from your social circle and that person will not exactly share your worldview. A translation will be required for you to fully understand each other. Pick a person from another culture and an even larger difference will separate your worldviews. The acronym WEIRD, standing for Western, Educated, Industrial, Rich, and Democratic, points out that the culture where 99% of science and scholarship takes place is in fact highly idiosyncratic compared to all the cultures of the world.[3] We are only beginning to comprehend the full diversity of cultural worldviews. Yet, it is possible to do so scientifically, once we recognize the need and set about studying it in the right way. An essential step is for worldwide cultural diversity to be represented within the scientific community.

Second, differences in worldviews *make* a difference in how a person behaves. In this respect, our worldviews are like our genes. In other words, each one of us is a collection of genes, called a genotype, that influence just about everything that can be measured about us, called our phenotype. Each of us is also a collection of beliefs, which goes by names such as worldview, meaning system, religion, ideology, or theory, that also influence just about anything that can be measured about us—the very same phenotype that is influenced by our genes. Recently, this comparison has been formalized from an evolutionary perspective as dual-inheritance theory,[4] in which worldviews are conceptualized explicitly as a stream of inheritance that first evolved by genetic evolution and has been co-evolving

---

[3] J. Henrich, *The WEIRDest People in the World: How the West Became Psychologically Peculiar and Particularly Prosperous.* New York: Farrar, Straus & Giroux, 2020.

[4] P.J. Richerson and R. Boyd, *Not by Genes Alone: How Culture Transformed Human Evolution.* Chicago: University of Chicago Press, 2005.

with it ever since, with the faster process taking a leading role in adapting people to their environments. To stress this comparison, the term "symbotype" can be added to the list of words that describe everything in our heads that influence our behaviors.[5]

Third, because differences in our symbotypes make a difference in our behaviors, there is a winnowing process. Symbotypes that cause us to behave dysfunctionally become less frequent over time. This winnowing process takes place over a wide range of time scales. At the time scale of genetic evolution, genes that resulted in dysfunctional symbotypes were selected against, leaving genes that endow us with impressive innate psychological machinery for forming and transmitting adaptive symbotypes. At the scale cultural evolution, most enduring symbotypes, evolving over centuries and millenia, impressively adapt human populations to their environments.[6] At the scale of an individual lifetime, both genetic and cultural evolution have endowed us with an impressive ability to assemble symbotypes that adapt us to our local environments.[7]

Fourth, people who become scientists bring their symbotypes with them. They will inevitably construct theories and hypotheses that "make sense" to them as everyday people—and defend them as ardently as most people defend their symbotypes, especially when threatened. This goes a long way toward explaining the "Sturm und Drang" of scientific controversies, which Denis and I recount in detail for the battles that we have fought.

* * *

[5] D.S. Wilson, S.C. Hayes, A. Biglan, and D. Embry, "Evolving the Future: Toward a Science of Intentional Change," *Behavioral and Brain Sciences*, 2014, **37**: 395–460.

[6] D.S. Wilson, *Darwin's Cathedral: Evolution, Religion and the Nature of Society. Chicago*: University of Chicago Press, 2002. See also D.S. Wilson, Y. Hartberg, I. Mac-Donald, J.A. Lanman, and H. Whitehouse, "The Nature of Religious Diversity: A Cultural Ecosystem Approach," *Religion, Brain and Behavior*, 2017, **7**: 134–153. [https://doi.org/10.1080/2153599X.2015.1132243.]

[7] B.F. Skinner, "Selection by Consequences," *Science*, 1981, **213**: 501–504.

Here's why I am passionate about defending Tinbergen's four questions against the suggestion that they have become obsolete: Because of my acute awareness that the Darwinian revolution is still a work in progress. The kind of transformation that took place for the study of ecology and behavior in the 1960s is still in progress for the human-related academic disciplines and has barely started for positive change efforts in real-world settings. As a quick way to make this point, imagine replacing the word "biology" in "Nothing in biology makes sense except in the light of evolution" with the words "human," "culture," or "policy," and imagine what kind of reaction you might get.[8]

Why the delay? In part because the study of evolution became so gene-centric during the twentieth century, in contrast to Darwin and other evolutionists of his day, such as Herbert Spencer. In part for the reasons that I just recounted. The closer scientific inquiry gets to studying the human condition, the more difficult it becomes for scientists to distance themselves from their everyday symbotypes.

While I engage in academic science, most of my effort is devoted to positive change efforts in real-world settings, such as a school for at-risk youth,[9] parks in disadvantaged neighborhoods,[10] methods of therapy and training,[11] and improving the organization and objectives of business groups.[12] Every major topic area (e.g., education, urban planning, business, health, environment) is a universe unto itself,

---

[8] D.S. Wilson, *This View of Life: Completing the Darwinian Revolution.* New York: Pantheon/Random House, 2019.

[9] D.S. Wilson, R.A. Kauffman, and M.S. Purdy, "A Program for At-risk High School Students Informed by Evolutionary Science," *PLoS ONE,* 2011, **6**(11): e27826. [https://doi.org/10.1371/journal.pone.0027826.]

[10] D.S. Wilson, "The Design Your Own Park Competition: Empowering Neighborhoods and Restoring Outdoor Play on a Citywide Scale," *American Journal of Play,* 2011, **3**: 538–551.

[11] S.C. Hayes, S.G. Hofmann, and D.S. Wilson, "Clinical Psychology is an Applied Evolutionary Science," *Clinical Psychology Review,* 2020, **81**: 101892. [https://doi.org/10.1016/J.CPR.2020.101892.]

[12] D.S. Wilson, M.M. Philip, I.F. MacDonald, P.W.B. Atkins, and K.M. Kniffin, "Core Design Principles for Nurturing Organization-Level Selection," *Scientific Reports,* 2020, **10**(1): 13989. [https://doi.org/10.1038/s41598-020-70632-8,]

inhabited by dozens of positive change methods. In the business world, there are thousands of consultants promoting their particular methods. The better known include Lean, Agile, and Sigma Six. They vary widely in their efficacy and assessment. Even the best are not provided with any kind of deep explanation for *why* they work. As a result, even the most effective positive change methods spread only to a degree and then come up against boundaries, like the geographical distribution of a species, beyond which they are unknown. And decades are required for them to spread to the extent that they do.

Against this background, a practical change method that can be given a deep evolutionary explanation, can be applied across all topic domains at multiple scales, with high standards of scientific research and assessment, is a big deal. It is nothing less than a catalysis of positive cultural evolution so that it can take place in a matter of years rather than decades or not at all.[13]

It is instructive to compare evolution, when presented as a practical change method, to the themes of my Dialogue with Denis. I begin by defining Darwinian evolution as any process that combines the three ingredients of variation, selection, and replication. This includes epigenetic and cultural evolution in addition to genetic evolution. Moreover, evolution can be an *intra*-generational process as well as an *inter*-generational process. The best example is the adaptive component of the immune system, but our open-ended behavioral flexibility—what B.F. Skinner called "selection by consequences"—qualifies as an intra-generational evolutionary process as well.[14] All these evolutionary processes can benefit from the powerful conceptual toolkit that was developed primarily in the context of genetic evolution.

Then, I make the crucial point that evolution doesn't make everything nice. It often results in outcomes that benefit me but not you,

---

[13] P.W.B. Atkins, D.S. Wilson, and S.C. Hayes, *Prosocial: Using Evolutionary Science to Build Productive, Equitable, and Collaborative Groups.* Oakland, CA: Context Press, 2019. (See also here.) [www.Prosocial.World.]

[14] B.F. Skinner, "Selection by Consequences," *Science,* 1981, **213**: 501–504.

us but not them, or short-term benefits at the expense of the longer view. It's not that virtuous behaviors such as cooperation, altruism, love, and a "seven-generation" mindset can never evolve--only that special conditions must be met. Positive change in real-group settings requires managing the process of evolution so that it becomes aligned with our valued goals. Otherwise, evolution will still take place but will result in problems rather than solutions. *Doing nothing is not an option.*

Next, the word "manage" requires examination. For many people, it implies the kind of top-down, centralized planning associated with socialist governments and hierarchical business corporations. This kind of management seldom works, because the world is too complex to be understood by any group of experts. Grand plans almost invariably succumb to unforeseen consequences. Also, whenever power is concentrated in the hands of a few elites, they almost always govern for their own benefit rather than for the benefit of the larger group. This is true even for socialist movements that started out well-meaning.

Instead, management must take the form of cautious experimentation. This experimentation is not random. Our partial knowledge does count for something, so we make educated guesses and test them against each other. Brainstorming often results in possibilities that seem to come out of the blue but then appear eminently sensible in retrospect. In addition to the experiments that we plan with rigorous methods such as randomized control trials and A/B testing, we should survey the "natural" variation that exists whenever different groups of people strive to do roughly the same thing independently of each other.

Once we identify a better practice, it is necessary to keep it going and replicate it at other locations. Replication tends to be taken for granted by evolutionary biologists because organisms just reproduce and their offspring develop into adults. None of this is guaranteed for a new cultural practice. The world is full of practices that function well but still fail to replicate. Also, a practice that works well in one context often must be modified to work well in other contexts, so cookie-cutter solutions are unlikely to work. Experimentation must be an ongoing process at all scales and contexts.

It is crucial to keep in mind that what counts as virtuous at lower scales can become disruptive at higher scales. Self-preservation is a good thing—until it results in self-dealing. Helping family and friends is a good thing—until it becomes nepotism and cronyism. Businesses maximizing profits for their shareholders is a good thing—until it leads to grotesque inequality. Growing a nation's economy is a good thing—until it overheats the earth. This multi-level dynamic reveals the necessity of a whole-earth ethic. We must plan our actions with the welfare of the whole planetary system in mind. First and foremost, we are human beings, citizens of the earth, and stewards for all other life forms. This primary social identity does not replace lower identities such as national, religious, or ethnic, but rather helps to orient them so that they become prosocial agents rather than disruptive agents within the larger global community.

In addition, the small group remains a fundamental unit of a society organized for the global good, like the cells of a multicellular organism. One of the most profound insights of modern evolutionary science is that our ancestors *never* lived alone. They *always* lived in the context of small groups engaged in the meaningful activity of collectively surviving and reproducing. This makes us more like ants in need of a colony than most people can fathom, especially against the background of individualism as the dominant intellectual tradition of the last 70 years.[15]

Our need to function in the context of small, appropriately structured groups engaged in meaningful work does not imply that individuals become mindless. On the contrary, the participation and checks and balances associated with democratic governance are part of the structure that is required to ensure cooperation and suppress

---

[15] D.S. Wilson and J.A. Coan, "Groups as Organisms: Implications for Therapy and Training," *Clinical Psychology Review*, 2021, **85**: 101987.

disruptive self-serving behaviors. By the same token, small groups are full participants in social organization at higher scales.

This narrative enables me and my colleagues to work with any kind of group, anywhere in the world, at multiple scales. It is intuitive and can be grasped by just about anyone with an inquiring and receptive mind. It portrays evolution as a process that can be intentional and needs to become much more so. Variation is portrayed as having both a directed and a blind component. Multilevel selection is foundational. Genes are included in the narrative, of course, but not as privileged entities. It is whole systems that need to evolve in all their complexity. As "wise managers of evolutionary processes" we need to understand the complexity of the systems that we design in great detail.

This narrative, which is designed to accomplish positive change in real-world settings, is fully consistent with the best of our current scientific knowledge about evolution—or so I claim. The only conflict is with claims about evolution that arose in the past—that it can have no purpose, that genes are privileged units, that higher-level selection is invariably weak—which need to be put to rest. I hope that my Dialogue with Denis helps to accelerate that process.

10. Denis Noble's Afterword

10.1. What Did I Learn?

I learnt a lot that is important from this Dialogue. That is a tribute both to the focused and civil Dialogue format sponsored by AcademicInfluence.com and to the constructive pleasure of interacting with David Sloan Wilson.

Perhaps that is not surprising. After all, we have both emphasised cooperation between living organisms as an important factor in evolution. David's book with Elliott Sober, *Unto Others: The Evolution and Psychology of Unselfish Behavior,* is a classic. I am also known as a critic of Richard Dawkins's "selfish gene" theory.[1]

It would probably have been very different in a Dialogue with one of the more dogmatic neo-Darwinists who have been very critical of my lectures and articles. I think that will have to come one day, but not quite yet. The reason is that the waters have been well and truly muddied by misunderstandings of my articles and lectures, and by confusions resulting from the bitter stand-off between creationists and evolutionists, particularly in the USA.

10.2. Misunderstandings

For around a decade, from the publication of *The Music of Life* in 2006 to the New Trends in Evolutionary Biology Meeting at The

---

[1] D. Noble, "Neo-Darwinism, the Modern Synthesis, and Selfish Genes: Are They of Use in Physiology?," *Journal of Physioogy,* 2011, **589**: 1007–1015.

Royal Society in 2016[2] and the publication of *Dance to the Tune of Life* the following year, I was the object of on-line ridicule and serious slander by some of the more dogmatic neo-Darwinists for having questioned the foundations of the Modern Synthesis.[3]

In a 2016 lecture[4] to The Physiological Society in the UK I reacted to a small but typical sample:

Misunderstanding 1: "Here we go again: someone arguing that DARWIN WAS RONG" (sic)

In truth, the people who argue that Darwin was wrong are precisely the dogmatic neo-Darwinists! They follow August August Weismann in saying that Darwin was wrong to include the inheritance of acquired characteristics and even more wrong to develop his theory of gemmules to explain it.[5] I argue not only that Darwin was right to do both of those and has now been vindicated (see **Darwin's gemmules** below), but that he was also right to argue for sexual selection being different from natural selection.[6] Julian Huxley's 1942 book *Evolution: The Modern Synthesis* does not even mention sexual selection. I have a far better claim to be a Darwinist than do the original neo-Darwinists.[7]

Misunderstanding 2: "His most moronic claim by far is the one on mutations not being random."

---

[2] P. Bateson, N. Cartwright, J. Dupré, K. Laland, and D. Noble, "New Trends in Evolutionary Biology: Biological, Philosophical and Social Science Perspectives," *Interface Focus*, 2017, **7**(5): 20170051. [http://doi.org/10.1098/rsfs.2017.0051.]

[3] D. Noble, "Evolution Beyond neo-Darwinism: A New Conceptual Framework," *Journal of Experimental Biology*, 2015, **218**: 7–13.

[4] [https://player.vimeo.com/video/194945801.]

[5] C. Darwin, *The Variation of Animals and Plants under Domestication.* London: John Murray, 1868.

[6] C. Darwin, *The Descent of Man, and Selection in Relation to Sex.* London: John Murray, 1871.

[7] D. Noble, "Charles Darwin, Jean-Baptiste Lamarck, and 21st Century Arguments on the Fundamentals of Biology," *Progress in Biophysics and Molecular Biology*, 2020, **153**: 1–4.

What I have said in my lectures and books is that randomness is *used* (harnessed) by organisms, so that, *after it has been used,* the outcome is far from random. It is not difficult to work this out, in particular from my contribution to the publication of the New Trends meeting, which is where I show how organisms can use stochasticity to produce non-random change[8] (see **Harnessing stochasticity** below).

Misunderstanding 3: "I know of not a single adaptation in organisms that is based on such environmentally-induced and non-genetic change."

The literature simply abounds in such examples. Whole books and extensive reviews are now devoted to transgenerational epigenetic inheritance.[9] Sure, there can be argument about the precise mechanisms and whether the Lamarckian interpretations are always correct, but there can no longer be argument about whether the Weismann Barrier has been breached. Regulatory RNAs have been shown to be transmitted to the germline[10] (see **Darwin's gemmules** below).

Misunderstanding 4: "Cells are transitory, and DNA is not"

Just the reverse is the case. My line of germ cells extends back through two billion years or so via reproduction of my ancestors to whatever were the first cells on earth. *My DNA does not!* It most probably did not even exist in the earliest cells. Anyway DNA is inactive without a cell. Only cells and the organisms they construct are alive.

---

[8] D. Noble, "Evolution Viewed from Physics, Physiology and Medicine," *Interface Focus,* 2017, **7**: 20160159.

[9] E. Danchin, A. Pocheville, O. Rey, B. Pujol, and S. Blanchet, "Epigenetically Facilitated Mutational Assimilation: Epigenetics as a Hub Within the Inclusive Evolutionary Synthesis," *Biological Reviews,* 2019, **74**: 259–282; doi.org/10.1111/brv.12453.

[10] D. Noble, "Exosomes, Gemmules, Pangenesis and Darwin," in L. Edelstein, J.R. Smythies, P.J. Quesenberry, and D. Noble, eds., *Exosomes: A Clinical Compendium.* Cambridge, MA: Academic Press/Elsevier, 2019; pp. 487–501.

Misunderstanding 5: The 2016 protest

These misunderstandings culminated in 2016 in the most important misunderstanding of all. I was the main organiser of a Discussion Meeting held jointly between The Royal Society (the UK national science academy) and The British Academy (the equivalent academy for the humanities and social sciences). The title of the meeting was "New Trends in Evolutionary Biology: Biological, Philosophical and Social Science Perspectives."

The meeting was held in November 2016, and the public announcement, inviting open registrations, was made 10 months earlier in February. The Royal Society regularly holds Discussion Meetings in all areas of science. Their purpose is to explore experiments and ideas in open discussion. There are many implications of theories of evolution to be found in philosophy and the social sciences, so you might imagine that this would be a welcome opportunity to explore those.

I was therefore deeply troubled to learn that immediately after the February public announcement, the President of The Royal Society received a protest letter from 21 distinguished scientists in the field of evolutionary biology. The protesters did not mince words. They wrote:

> It is hardly an exaggeration to say that the situation is similar to the Society allowing advocates of homeopathy to organize a meeting on medical research.

This was directed at me by name. The other four distinguished organisers—two scientists and two philosophers—were not named. The slur about homeopathy was, of course, ridiculous. Nothing in what I have written could possibly justify it.

We reacted by explaining that we welcomed other views, had invited standard neo-Darwinists to the meeting and were open to add more. In the course of negotiations over the matter it also became clear to me that none of the signatories to the protest could have read my books or articles. It is an absolutely fundamental professional procedure in academic life to read an author before attacking their work. Then it is perfectly reasonable to do so openly and in print. Open discussion and debate are the lifeblood of science.

The meeting eventually went ahead as planned. It resulted in a valuable publication of around 20 articles published by The Royal Society in 2017.[11]

10.3. Peace After the Storm

Since 2016 these kinds of absurdly wide-of-the-mark comments have stopped. It is a major advance that discussion and debate in the field have become more civilised. The protest received by the Royal Society achieved a very important, but completely unintended, outcome. The protest over-reached itself and so reflected the ridicule back on the protesters.

I celebrate the fact that arguments in evolutionary biology can be courteous and respectful. This Dialogue contributes enormously to that development, which is a tribute to the Editor's vision.

10.4. What is Meant by Neo-Darwinism?

During the Dialogue, I learnt that many, like David, who use the term neo-Darwinist, are really what I would call "modern Darwinists"; they are Darwinists updated (that being their sense of "neo-"). In David Sloan Wilson's case, the update is achieved by defining it in terms of Tinbergen's four questions, rather than by focussing, as I do, on the Weismann Barrier. All I will note here is that this is not the sense of "neo-" that was introduced by G.J. Romanes in 1883.[12] I believe that Romanes was concerned to defend Darwin's distinc-

---

[11] P. Bateson, N. Cartwright, J. Dupré, K. Laland, and D. Noble, "New Trends in Evolutionary Biology: Biological, Philosophical and Social Science Perspectives," *Interface Focus,* 2017, **7**(5): 20170051. [http://doi.org/10.1098/rsfs.2017.0051.]

[12] G. Romanes, "Letter," *Nature,* 1883, **27**, 528–529.

tion between artificial and natural selection,[13] and his adherence to Lamarckian forms of inheritance.[14] But, I readily acknowledge that the general sense of "neo-" is "new." Wikipedia notes that "it can mean any new Darwinian- and Mendelian-based theory." I fully understand why David and some other biologists use the term in this way.

10.5. Confusions

There is a significant difference between the way in which religion and science interacted over the theory of evolution in North America and in many other countries. In Europe the two largely made their peace long ago. Even in the nineteenth century there were leading clerics (including an Archbishop of Canterbury) whose religious concepts, particularly their theology, posed no problems for them accepting what Darwin had shown, and that is certainly true today.[15] That is even more true of many Oriental religions. Even the concept of a god in many religions is far from the creationist concept. Some, such as Buddhism, can even be interpreted as a form of atheism. The Buddha is not venerated as a god.

He is venerated for his insights on humanity and suffering, and for the healing meditative tradition that he founded.

I believe therefore that it is extremely unfortunate that the debate on evolution became so polarized between Christian fundamentalists and evolutionary biologists, particularly in the USA, where 40%

---

[13] C. Darwin, *The Descent of Man, and Selection in Relation to Sex.* London: John Murray, 1871.

[14] C. Darwin, *The Variation of Animals and Plants under Domestication.* London: John Murray, 1868.

[15] "Rev. Frederick Temple (1821–1902), preached a sermon at the British Association for the Advancement of Science in Oxford showing his appreciation of *The Origin of Species.* He epitomised the learned and liberal Anglican and became Archbishop of Canterbury in 1896. He gave the Bampton Lectures on *The Relations between Religion and Science* in 1884. Temple had a good understanding of contemporary science and out of his eight lectures, two were affirmative of evolution. He discussed the creation accounts of Genesis which he saw as allegory," cited in article on website of British priest-geologist, Michael Roberts, "Peddling and Scaling God and Darwin." [https://michaelroberts4004.wordpress.com/2015/02/23/evolution-and-religion-in-britain-from-1859-to-2013/.]

of the population adhere to young-earth creationism, i.e., that the earth is less than 10,000 years old and that species were separately created.[16]

The reason I raise this issue is that I also think that the polarization has made it difficult for the more dogmatic neo-Darwinists to react reasonably to fundamental criticisms of the Modern Synthesis from other scientists. Their fear may be that any doubt about the Modern Synthesis will be seized on by the creationists. That fear is not without foundation. Some of the websites featuring (pirated!) videos of my lectures, particularly that in Suzhou, China, in 2012,[17] may be found on creationist or intelligent design (ID) websites. All I want to note here is that it is a serious misunderstanding of my lectures to interpret them as supporting creationist or ID arguments. My interpretation of purpose (teleology) is that it evolved and that it is now itself a driver of evolution. Evolution itself evolves new processes by which it occurs. Organisms themselves bring purposiveness to the universe.

10.6. What Has Developed Since the Original Dialogue?

I identify three developments that seem to me to be ground-breaking: (1) a historical discovery on the origin of the Tree of Life idea; (2) the identification of extracellular vesicles as Darwin's gemmules; and (3) the harnessing of stochasticity.

10.6.1. The Tree of Life

Charles Darwin is justly famous for his discovery in his 1837 "B notebook" of the Tree of Life. He roughly sketched his idea based

---

[16] See here. [https://news.gallup.com/poll/261680/americans-believe-creationism. aspx.] It is also important to note that the US is not alone. Other strongly creationist countries include Saudi Arabia (75%), Turkey (60%), Indonesia (57%), Brazil (47%), and Russia (34%)—see here. [https://ncse.ngo/polling-creationism-and-evolution-around-world.]

[17] [http://www.voicesfromoxford.org/physiology-and-the-revolution-in-evolutionary-biology/184/.]

on the radiation of the species of finch that he discovered in the Galapagos Islands. This is a cornerstone of modern evolutionary biology and is often used to contrast the evolutionary views of Darwin and Lamarck. After the Dialogue, I reread Lamarck's work more carefully. I happen to have fluent French so I did so in his own language. I was astonished to find an even more detailed tree of life at the very end of Lamarck's 1809 *Philosophie zoologique*.[18] This is so important that I reproduce the diagram here.

Lamarck's diagram not only predates Darwin's famous "B" notebook diagram by 28 years, it is also specific about which large animal groups developed from which, and how they branched. The difference in the two tree diagrams lies in the range of species to which they are applied: to the Galapagos finches in Darwin's notebook tree; and to a different and wider range of life forms in Lamarck's tree. Yet modern textbooks on evolution still claim that "Darwin's conception of the course of evolution is profoundly different from Lamarck's in which the concept of common ancestry plays no part."[20] What I found in Lamarck's 1809 book shows that this is simply incorrect.

I was so excited by this discovery that I initially thought that I must have made a new historical discovery. That is the extent to which many textbooks on evolution have misled me and many others! Lamarck's concept of the Tree of Life is not recorded in any of them. But, of course, it did not make sense that this obvious fact in Lamarck's work had gone completely unnoticed. It is acknowledged in several publications by historians of science.[21] Moreover, Stephen

---

[18] J.-B. Lamarck, *Philosophie zoologique*, introduced and annotated by André Pichot. Paris: Flammarion, 1994; this is the text of the original 1809 edition.

[20] D. Futuyma and M. Kirkpatrick, *Evolution*. Oxford University Press, 2017; p. 13.

[21] J.D. Archibald, "Edward Hitchcock's pre-Darwinian (1840) "Tree of Life," *Journal of the History of Biology*, 2009, **42**: 561–592; P.J. Bowler, *Evolution: The History of an Idea*. Berkeley, CA: University of California Press, 1984; N.P. Hellstrom, "Darwin and the Tree of Life: The Roots of the Evolutionary Tree," *Archives of Natural History*, 2012, **39**: 234–252; K.K. Misra, "*Philosophie zoologique*—200: Lamarck in Retrospect," *Science as Culture*, 2011, **77**: 198–207; M.F. Oxenham, "Lamarck on Species and Evolution," in A.M. Behie and M.F. Oxenham, eds., *Taxonomic Tapestries: The Threads of Evolutionary: Behavioural and Conservation Research*. Canberra: Australia National University Press, 2015; pp. 155–

Jay Gould in one of his last books[22] not only refers to Lamarck's tree of life, he comments:

> How can we view his [Lamarck's] slow acknowledgement of logical error, and his willingness to construct an entirely new and contrary explanation, as anything other than a heroic act, worthy of our greatest admiration and identifying Lamarck as one of the finest intellects in the history of biology?

Enough said.

Of course, the idea of the Tree of Life has in any case now become a Network of Life. Lateral interactions between species occur as well as vertical inheritance.

10.6.2. Darwin's Gemmules

Even a casual reading of *The Origin of Species* shows that it freely assumes the in- heritance of acquired characteristics. Mayr, in his magisterial book,[23] identifies 12 places where this is the case. This issue was one that fed Darwin's doubts and caution for many years. In his 1868 book,[24] he formulated his theory of gemmules, in which he solves his problem in almost exactly the same way as Lamarck did. Puzzled by the same question, which is how information from the soma could be transmitted to the germ-line, he postulated the existence of invisible "subtle fluids." Darwin's gemmules perform the same function as Lamarck's fluids. The difference is that Lamarck used the field concept, as for example in electromagnetism. Darwin used a particle concept.

170; see, also, J.-B. Lamarck, *Philosophie zoologique,* introduced and annotated by André Pichot. Paris: Flammarion, 1994.

[22] S.J. Gould, "A Tree Grows in Paris: Lamarck's Division of Worms and Revision of Nature," in *idem, The Lying Stones of Marrakech.* New York: Harmony Books, 2000; pp. 115–143.

[23] E. Mayr, *The Growth of Biological Thought.* Cambridge, MA: Harvard University Press, 1982.

[24] C. Darwin, *The Variation of Animals and Plants under Domestication.* London: John Murray, 1868.

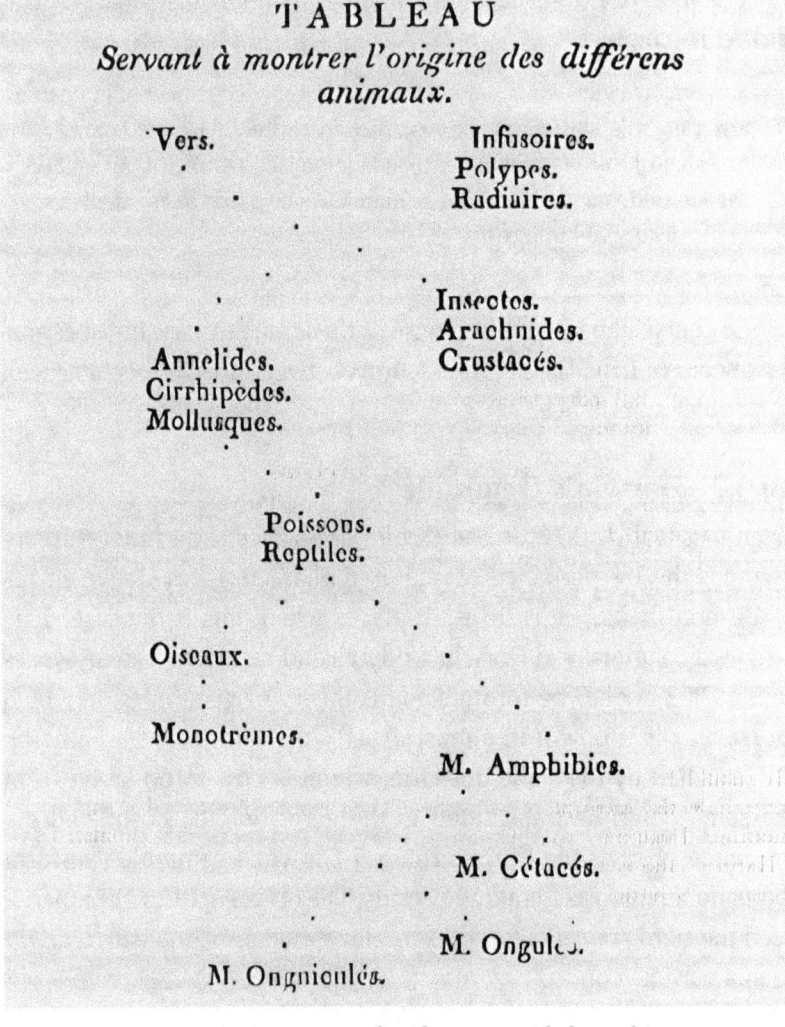

Figure 10.1: **Lamarck's Tree of Life.** From *Philosophie zoologique*, 1809. The root of the tree is at the top (*vers* = worms). There are then several branchings leading to many different kinds of animals. The edition in which I first saw this tree diagram was the Flammarion reprint of 1994 in which the dotted lines have been replaced with full lines. This version is taken from Voss (1952)[19] and originally copied from the 1830 reprint. I use this version as a tribute to Voss's careful historical study of keys and trees in evolutionary biology.

But do those particles exist? This is a question as big as the one that followed the discovery of the circulation by William Harvey. That question was solved by the development of the light microscope and the discovery of microscopic capillaries. We have had to wait a long time for light microscopy to achieve the much higher resolution needed to view what are almost certainly Darwin's postulated gemmules. They have now been seen, they are numerous, and they are called "exosomes." These are tiny extracellular lipid vesicles (EVs) packed with RNAs, DNAs and other molecules that contain information on the regulatory state of the genome from the cells that extruded them. It has taken a remarkable 10-fold increase in the resolution of light microscopy to visualize them.[25]

Just as Darwin postulated, exosomes can transmit their RNAs, DNAs, and other molecules to the germ-line, so crossing the Weismann Barrier, which was supposed by him and the founders of the Modern Synthesis to prevent precisely this transmission from happening. This is a profound break from the fundamentals of the Modern Synthesis. It is therefore important now to discover experimentally what transgenerational effects can be attributed to exosome uptake by the germline. There is a whole new field of research rapidly opening up here. The implications for the inheritance of disease states are important.[26]

[19] E.G. Voss, "The History of Keys and Phylogenetic Trees in Systematic Biology," *Journal of the Scientific Laboratories, Denison University,* 1952, **43**: 1–25.

[25] L. Edelstein, J. Smythies, P. Quesenberry, and D. Noble, eds., *Exosomes: A Clinical Compendium.* Cambridge, MA: Academic Press/Elsevier, 2019.

[26] See, e.g., P.D. Gluckman and M.A. Hanson, "Developmental Origins of Disease Paradigm: A Mechanistic and Evolutionary Perspective," *Pediatric Research,* 2004, **56**: 311–317; doi: 10.1203/01.PDR.0000135998.08025.FB; and Sandra M. Rehan, Karl M. Glastad, Michael A. Steffen, Cameron R. Fay, Brendan G. Hunt, Amy L. Toth, "Conserved Genes Underlie Phenotypic Plasticity in an Incipiently Social Bee," *Genome Biology and Evolution,* 2018, **10**: 2749–2758. [https://academic.oup.com/gbe/article/10/10/2749/5106031.]

10.6.3. The Harnessing of Stochasticity

The misunderstanding about whether mutations are or are not random is central to what I have written on this question.

The 1942 Modern Synthesis portrays organisms as passively experiencing mutations. I turn that idea on its head. Stochasticity is used *functionally* by organisms to create novelty that is *directed*. I first outlined that idea in 2017 in an article[27] forming part of the special issue of *Interface Focus* arising from the New Trends meeting, where I wrote:

> *A central thesis of this paper is that blind stochasticity is a misconceived idea as it has been used in evolutionary biology. Stochasticity is used by organisms to generate new functional responses to environmental challenges. Far from proving that evolution is necessarily blind, randomness is the clay from which higher-level order can be crafted. But it necessarily works the other way too: higher levels then organise the molecular level through many forms of constraint.*

There is nothing strange about the harnessing of stochasticity. All physiological control systems canalize random events at a micro-level to create order at a macro-level. Plants use absorption of the random arrival of photons to create energy that is then used to create and maintain structural order. The immune system uses random variations in the immunoglobulin sequences to be selected for functionality as antibodies. The nervous system uses stochasticity at all levels to generate anticipatory behaviour.[28] Importantly, this includes the ability to act rationally and with purpose.[29] It is therefore now possible to provide physiological explanations for purposive behaviour in organisms. Darwin was correct to highlight the existence of

---

[27] D. Noble, "Evolution Viewed from Physics, Physiology and Medicine," *Interface Focus*, 2017, **7**: 20160159.

[28] R. Noble and D. Noble, "Harnessing Stochasticity: How Do Organisms Make Choices?," *Chaos*, 2018, **28**: 106309.

[29] R. Noble and D. Noble, "Can Reasons and Values Influence Action: How Might Intentional Agency Work Physiologically?," *Journal of the General Philosophy of Science*, 2021, **52**: 277–295.

such behaviour when he wrote in his 1871 book: [30]

"Just as man can give beauty [in his breeding of animals] ....so it appears that female birds in a state of nature, have by a long process of selection of the more attractive males, added to their beauty."

On the previous page, he makes it clear he means *conscious* choice:

...and consciously exert their mental and bodily powers.

10.7. The Future

Research on evolution is clearly alive and well, and continues to spring surprises onto the world of science. The Third Way of Evolution,[31] a group I helped to form with James Shapiro, has now identified around 40 major discoveries that lie outside the original 1942 framework of the Modern Synthesis. Extremely few of these are even mentioned in textbooks and popularizations of evolutionary biology. Some are extensions, but some are incompatible with the original Modern Synthesis. There are many opportunities here for further research.

I hope that will also be the outcome of this Dialogue.

---

[30] C. Darwin, *The Descent of Man, and Selection in Relation to Sex.* London: Penguin Classics, 2004; p. 246. (First edition: London: John Murray, 1871.)

[31] [www.thethirdwayofevolution.com.]

Index

www.ingramcontent.com/pod-product-compliance
Lightning Source LLC
Chambersburg PA
CBHW060129130626
46556CB00006B/2280